NO LONGER THE PROPERTY OF
LONGMONT PUBLIC LIBRARY

AF693321

629.133133 LEV
Levin, Kristine Cox, 1944-
Silent wings :
experimenting with paper

1000

SILENT WINGS

For the Children at the Open Living School
who were both
my students and my teachers

629.133133

SILENT WINGS

Experimenting With Paper Airplanes and Balsa Gliders

By Kristine Cox Levin

OUT OF PRINT

PRUETT PUBLISHING COMPANY
Boulder, Colorado

01764
LONGMONT PUBLIC LIBRARY

© 1977 by Kristine Cox Levin

All rights reserved, including those to reproduce this book, or parts thereof, in any form without permission in writing from the Publisher.

First Edition

1 2 3 4 5 6 7 8 9

Printed in the U.S.A.

Library of Congress Cataloging in Publication Data
Levin, Kristine Cox, 1944-
Silent wings.

Includes bibliographies.
1. Paper airplanes. 2. Gliders (Aeronautics) — Models., 3. Aerodynamics. I. Title.
TL770.L47 629.133'1'33 75-17908
ISBN 0-87108-193-8

Illustrations were drawn by the author and Susan Wellman

Acknowledgments

I have a warm feeling of appreciation for several people who were very supportive of me as I was working on this, essentially very basic, study. In both direct and indirect ways I found that my efforts were met with genuine interest and helpfulness from numerous colleagues and acquaintances as well as from dear friends of long standing.

To Judy Litschel, I say thanks for the support and inspiration and appreciation she offered while I worked in her classroom, which was a very effective, positive learning environment for the children. To Dr. Bob Christopher, I say thanks for the many hours of personal attention, and interest in our group. He enthusiastically taught us the best methods for building with balsa as well as offering information about flying conditions and resources for our further exploration of this topic.

My own personal interest and delight in air and flight was greatly heightened by my studies with Dusty Blades in his class. His unusually clear explanations of aerodynamics helped me understand and express my understandings better, to the children.

To Patti McKinnell and Jean Jacobsen, I say thanks for reading over the manuscript and pointing out directions (for making paper airplanes) that were unclear, as well as those that were clear, and for expressing such enthusiasm about the project in general.

My appreciation goes to Barry Kluger for reading over the manuscript also, and for offering his model airplane and the air siphon suggestion to be included.

I have a special thanks for Dr. David Hawkins and Frances Hawkins. I thank Dr. Hawkins for his specific suggestions of ways to study air (such as swishing the air bubble back and forth) and I thank the two of them for that greater support they've offered me as a student and teacher, learning about the significance of play and much more.

And finally, and fully, of course, the children. . . my students in one way, and my teachers in other ways. The eight, nine, and ten year old children, who were active in the Glider Club at Edgewater Open Living School, Edgewater, Colorado, are a main motivation for this book. There were some children who maintained an interest in gliders all that year and even the following years: Pam Reid, Mary Brainerd, Brian Monroney, Bradley Huntting, Brenda Mayes, Danny Byars, and Chuck Hopper. Most of them have taught other children what they know, and have pursued their interest at home and in other classrooms, at school. I am delighted that I could include the models they made. Although this handbook is mostly about the information that we learned about gliders, for me it is also a memory of the relationships and experiences that we shared together, which I enjoyed so much.

Kristine Cox Levin
Boulder, Colorado 1975

A Note about the Teaching Method

Our paper airplane study, which later grew to include the balsa glider work, and study of air, was a result of my efforts as a teacher to deepen and expand an interest that the children had in paper airplanes. I felt some fascination for the airplanes myself, and so began exploring ways to give children more success and variety of experience with the airplanes and ways which would offer them a deeper and more satisfying involvement in the subject. The Glider Club was what the group of children called themselves. The club grew from their interests and studies in the classroom. A picture of the Glider Club as it functioned within the classroom setting would include: a small group of children going on trips to lumber yards to buy balsa, discussing the cost, adding and dividing to obtain an average for everyone to chip in and buy the wood, back to the classroom to cut and transfer patterns, discuss exacto knives, (pro, cons, costs and qualities) build, sand, glue the balsa, balance planes, test flights, and discuss the control surfaces, and airfoils.
The airfoils proved to be an exciting separate study for a while as we studied lift and did a lot of studies of air. Just as we were finished with the glider study there was an article in *Time*, April, 1973, p. 51-52 about a man from Georgia who has explored paper airplane designs and found one that has 17% more lift with an upside-down airfoil!

The club spent a lot of time doing science demonstrations and figuring out why they worked or didn't work. Dr. Christopher came to talk with us several times, demonstrating not only the building and flying of gliders, but also explaining about air currents, thermals, balancing planes, and correcting and adjusting flights which we made out on the field nearby. We gathered to talk and hear stories as well, about the early flights, so frightening and yet exciting, taken by the pioneers of air travel. The evaluation process was continued and ongoing on the part of the children as they built their gliders, and asked the ever present question "why?" to understand better, the reason for the glider performances. In seeking answers to their questions and studying air, and then reapplying the information they had learned, they were frequently able to correct the difficulties they had encountered with minimal adult guidance in many cases. At the same time, as a teacher I was observing and evaluating how much information each of the children understood during our discussions and what experiences should follow next to help them understand the phenomena that was pertinent at the time.

Since that year many of the same children have been interested in gliders intermittently. They have continued to make gliders at home, and have initiated further studies of flight and paper and balsa glider work in school. They have taught other children what they know and have drawn others into the study. This was a confirmation for me, of the depth of our study and of the understandings most of them had acquired. We had worked a great deal with materials and concrete objects and field experiences so that the learning would be as full and meaningful as possible. Then in our discussions we reflected, pondered and experimented with ideas. We were relating the theory and information we found out to the real world and daily living.

My commitment as a teacher, like that of many teachers, is to a humanistic approach to education, in which I teach the child and relate to the child as a whole person. This focus encourages her or his autonomy and independence as a responsible person (and ultimately citizen) and as a discriminating and thoughtful student. So the child's emotional and social development is equally as important as her or his skill and academic progress. This means for me an individualized classroom where children are challenged according to their abilities and their learning is centered on meaningful experiences which they can relate to.

Our work with academic skills is reflected in this handbook also. In teaching skills in ways that would promote their positive self-concepts I used and encouraged the children's writing abilities heavily. The following is the method I used: First, the children are always encouraged to write anything they are interested in, and are given a lot of individual attention and adult help for this activity. Then, children are gathered in small groups, or individually, to work on skills or specific problems they might have in doing their work. As a teacher I avoid making corrections on the papers they write. Instead I note which errors are most frequent and have a skill lesson on those things. During the skill lesson, the child's attention is focussed on the skill needed for his or her work, but not on the ideas he or she is trying to express in the work. Positive feelings are maintained toward both subjects: skills and writing, as well as toward the self. The skill is not negative interruption of something the child is trying to do or a useless boring lesson with a whole group, on something the child doesn't need. The writing is not discouraging to try to do for fear of being wrong or of having red correction marks on one's own work. And the child feels capable of learning and remembering the information he is learning.

This study was tremendously rewarding for me as a teacher, like the indepth work that many of my colleagues have done. I have wanted to share this, and teach it and encourage other teachers to explore it too. Yet, I know I can't "give" my experience to others.

So I hope to offer some information in the handbook that can be used as a beginning and springboard for readers who want to explore this topic. I would hope that at best, the book will be used as a catalyst and resource for people to further pursue their interests in air and air vehicles. And, that people will be actively involved as they read this, by making the airplanes and doing the experiments!

Contents

This book about airplanes has the freshness of discovery about it. Kris Levin didn't invent the airplane, but she invented ways for herself and her children to investigate airplanes and make them fly. When teachers do this kind of thing, and involve their children and readers in it, everybody gets airborne.

David Hawkins

Part One

Building Paper Airplanes

Ode to a Paper Airplane

Airy and white
silently grazing desk rows
you zip or float
and then alight,
gracefully perched at her toes.

She masks her surprise
pausing with chalk in her hand,
then carries on
in secret delight
marking how smoothly you land.

For years humankind
lightly has treated your charm;
though childrens' eyes
have followed your flight,
In schools you are viewed with alarm.

Let me just say
as you silently sail through the sky,
with slow gentle glide,
new times begin today.
I think your role will change by and by.

What Is A Glider?

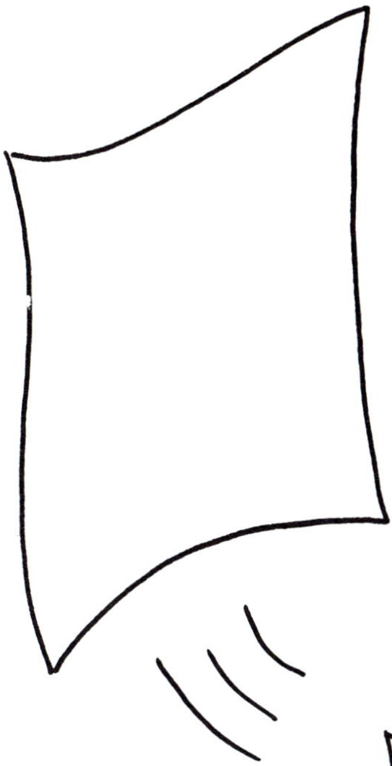

Do you have a few pieces of paper handy? Take a piece of paper, and let it fall to the floor. You see that in a way, it glides or flies.

But it doesn't fly straight and smooth. It darts back and forth, going one way and then another.

This darting and swooping happens because the paper is UNSTABLE.

There are many ways to make it STABLE. When this is done, the plain sheet of paper becomes a glider.

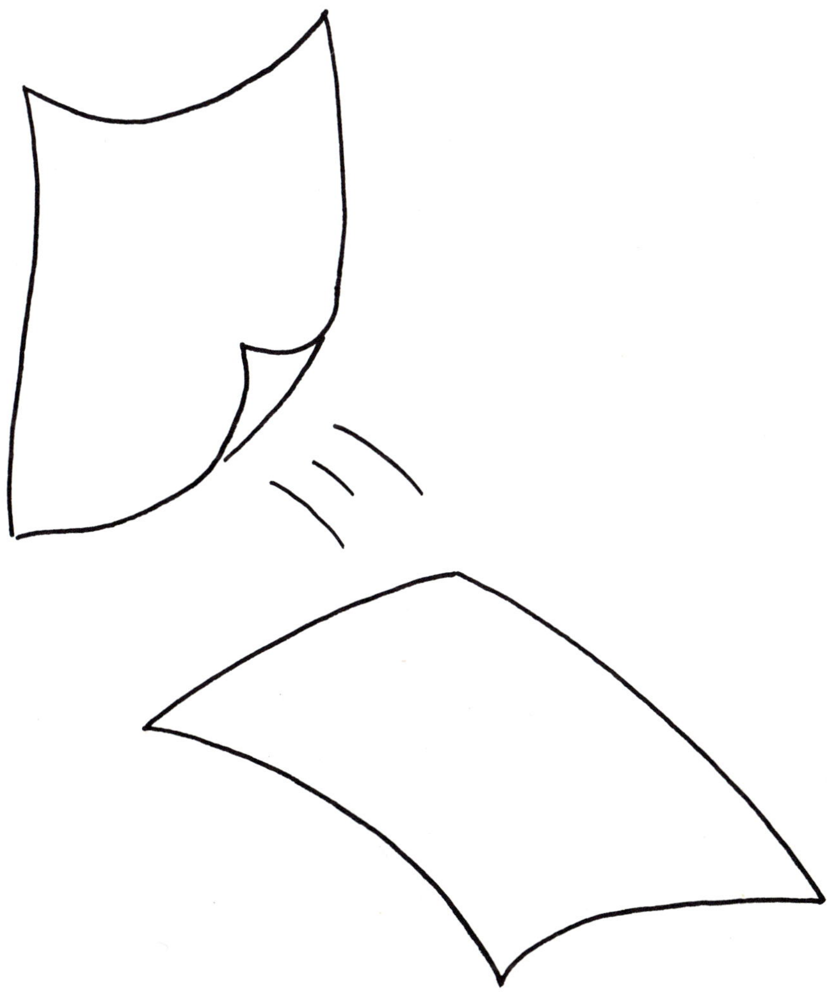

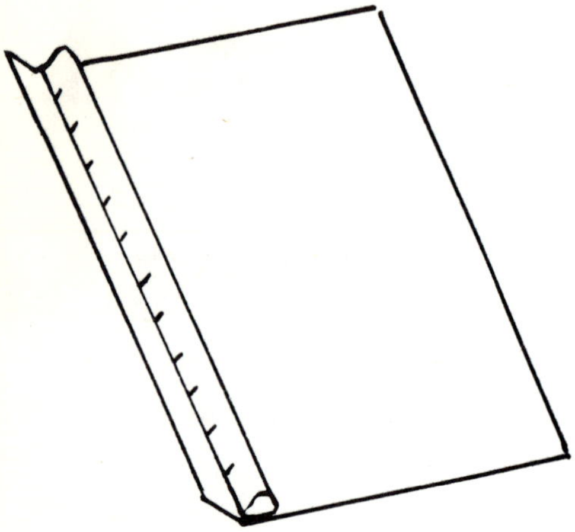

One way to make it STABLE is to fold the long side of paper over itself in a narrow fold (about ½ inch or ¾ inch) four or five times. Fold it against the side of a ruler to help get the folds straight.

The Leading Edge

Then crease the folded edge (called the LEADING EDGE) in the center. Now LAUNCH it. Try launching it gently, then firmly. Aim it nose up, nose down, or level. Experiment as you throw, to find out which way it flies the best.

Another way to do the same thing is to place paper clips on the LEADING EDGE.

Add enough clips to give it a steady, smooth glide path.

Trailing Edge

Try holding the plane by the nose when you launch. See if this is different. Is it better or worse than holding it by the trailing edge (the back of the plane.)

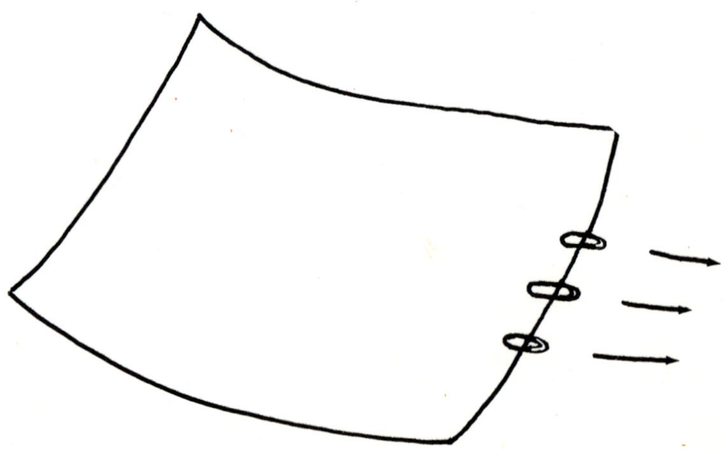

Here is another way to fold your paper for a more stable flight. It is the most commonly made paper airplane, frequently called a dart.

Be sure to make the edges meet and crease each fold with your finger or thumb.

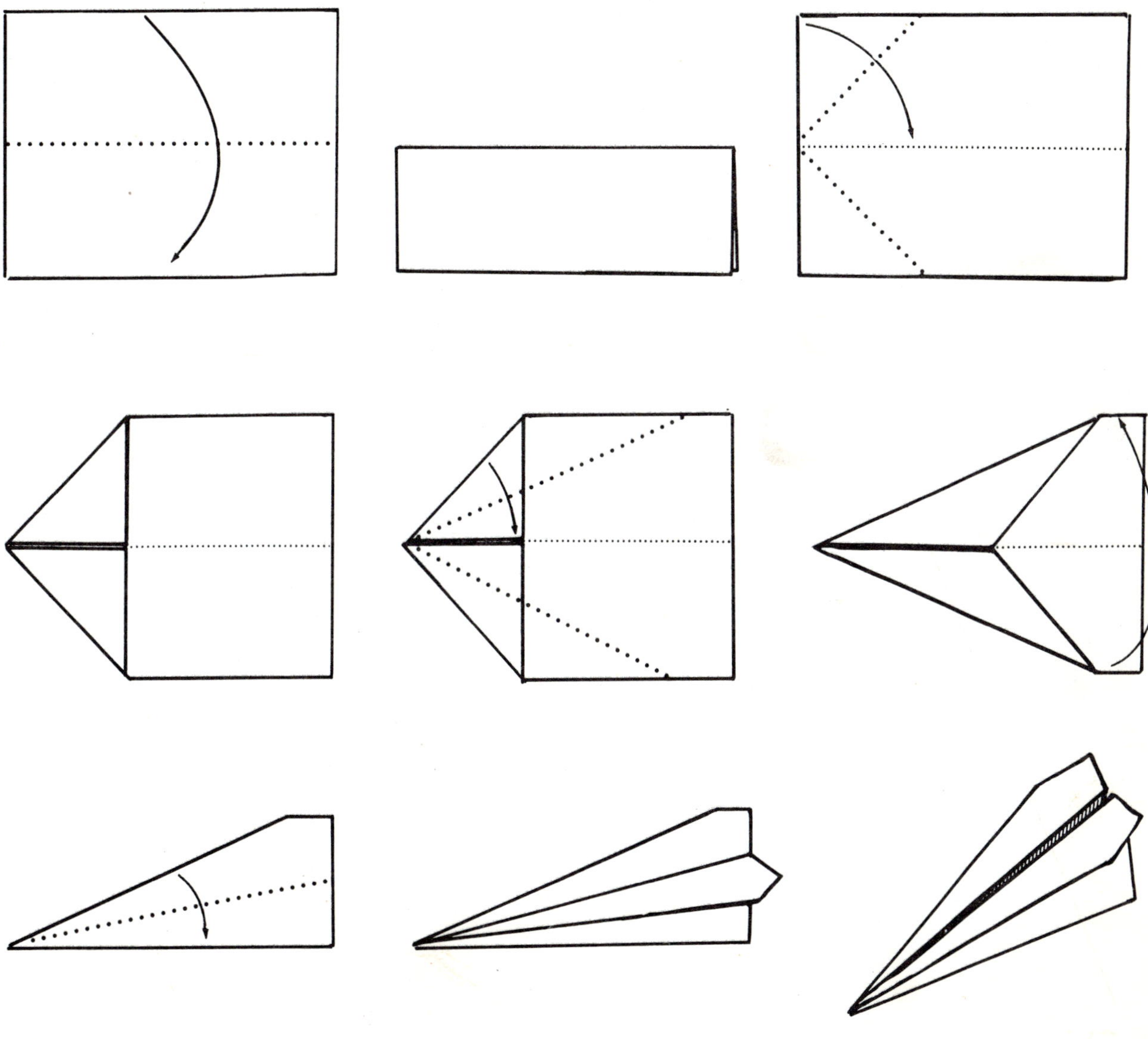

Launching Your Glider

Now LAUNCH your dart. The launch makes a big difference. Try aiming it nose up, nose down, or nose level. Throw it hard, then very gently. Experiment until you find the best way for your airplane to be launched.

Try folding back the nose of the dart. You can do it with almost any paper glider you might invent and it's a good way of adding weight to the nose end for stability and balance.

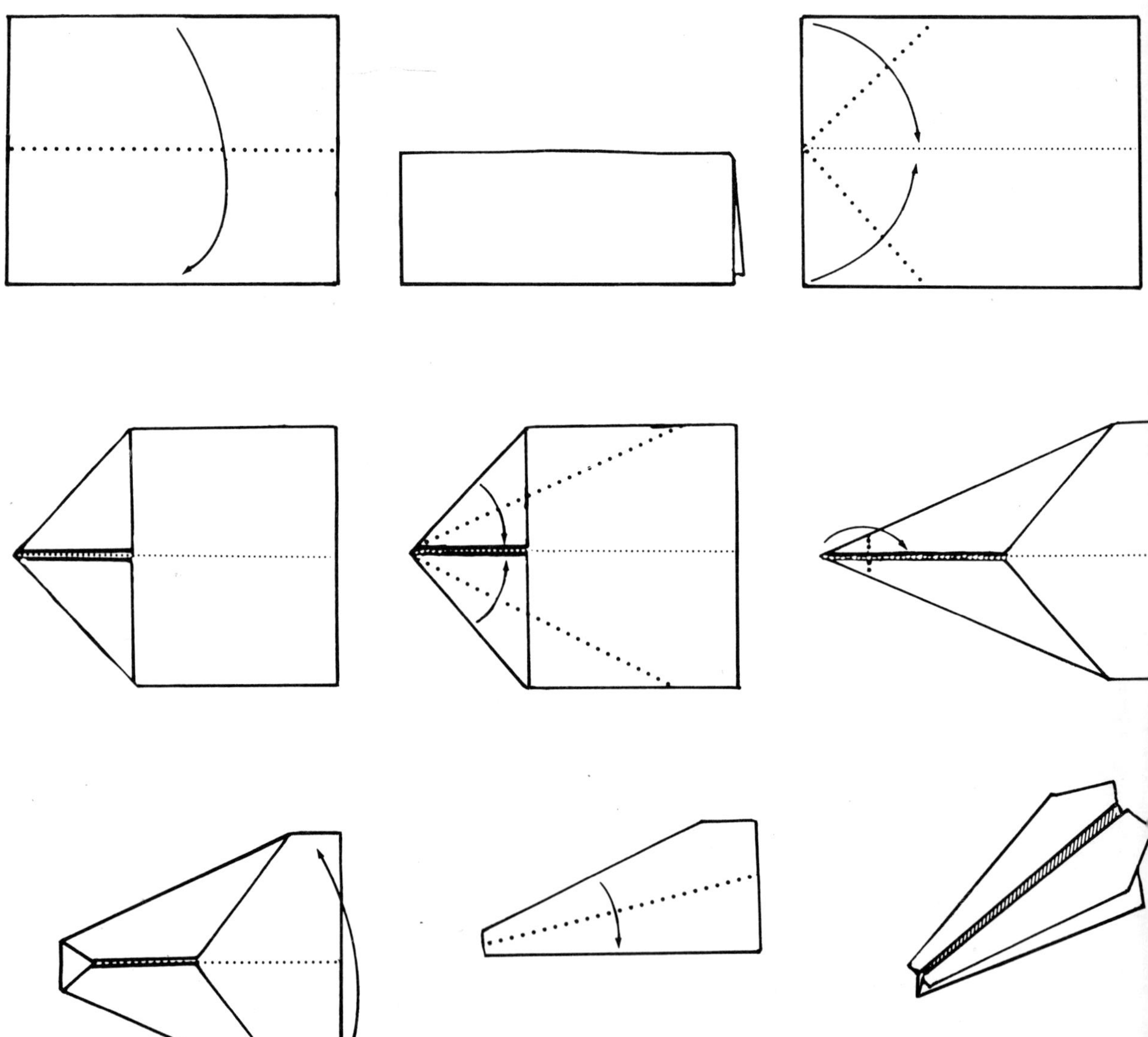

The Stall

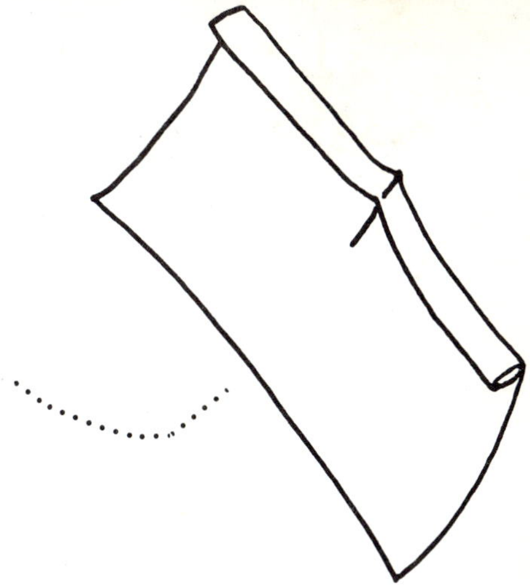

If your glider STALLS (which means the nose goes up and then suddenly the glider swoops down) that means you will need to add weight to the nose, or take weight off of the tail.

Add some more folds of paper, or add some clay, paper clips, alumi-unm foil, erasers, staples, or anything you think will work.

You may also want to try a different launch. Hold it by the nose to launch it. Try a gentle launch. Try a slower launch, aiming the nose more level or slightly down. It could STALL from being launched too fast, or launched nose up, too much.

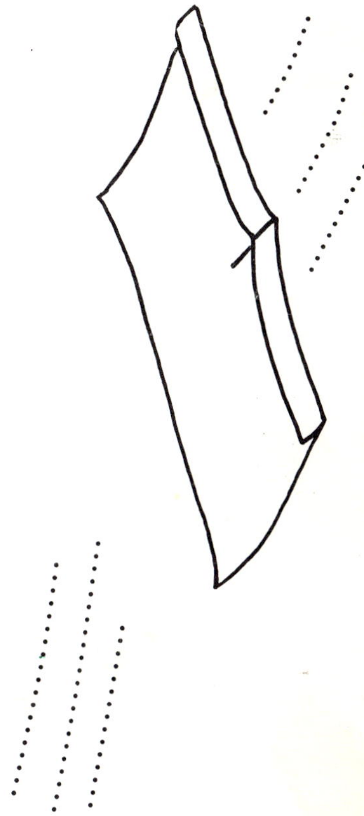

The Nose Dive

If your paper glider does a NOSE DIVE, maybe you are aiming it with the nose down too much when you launch.

But a NOSE DIVE can also mean you need to add weight to the tail, or take weight off of the nose.

You can remove a paper clip, or whatever you've added. If you have added nothing and it still nose dives, you'll need to add weight to the tail. See p. 9.

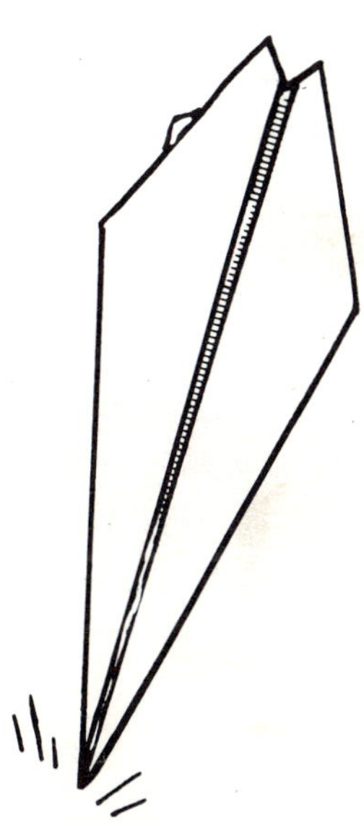

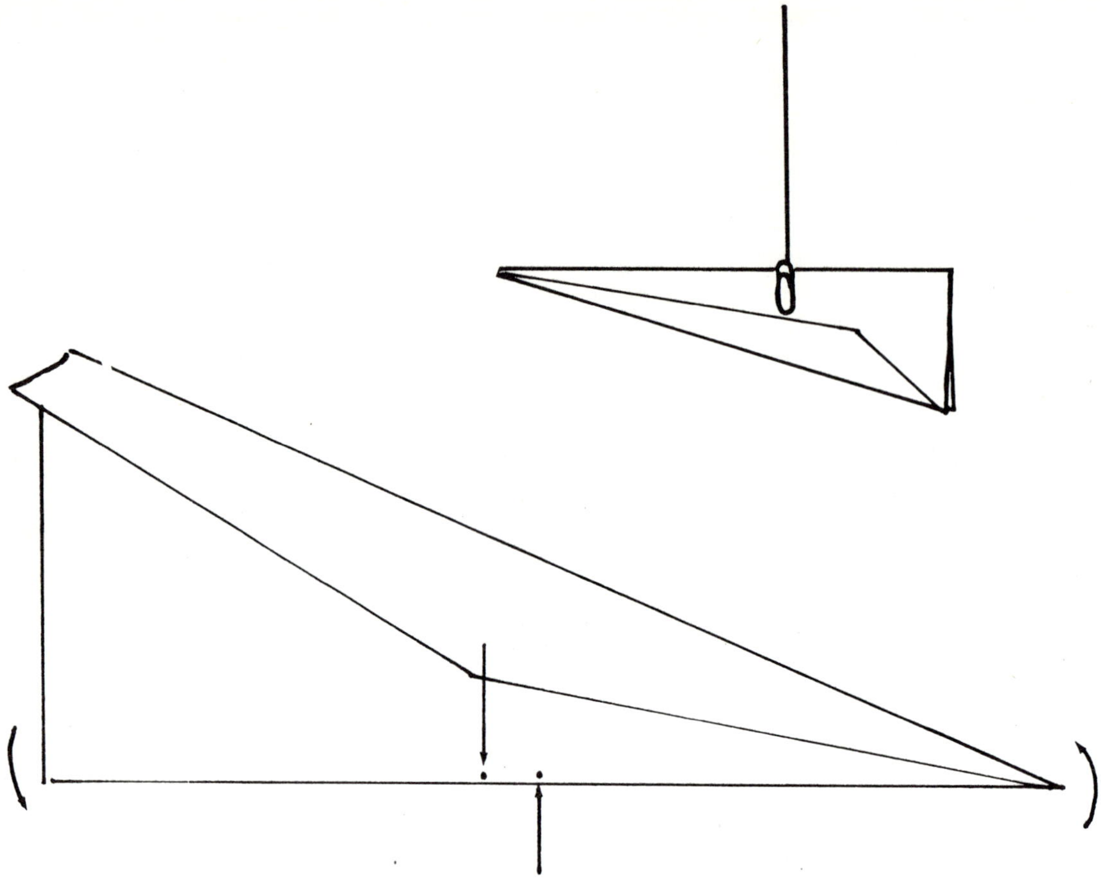

Center of Gravity and Center of Lift

The reason you can correct a STALL or DIVE this way has to do with the locations of the CENTER OF GRAVITY and the CENTER OF LIFT.

The CENTER OF GRAVITY

Or (C.G.) is the balance point of forces pulling the plane down. You can find this balance point by balancing the plane upside down on your fingers and marking the point, or balance it, hanging from a string on a paper clip and mark it.

The Center of Lift (or C.L.) is that point where the forces holding the plane up, are balanced.

If the C.G. is behind the C.L. the plane will nose up. So, by adding weight to the nose, you move the C.G. forward and it won't nose up. Doesn't this remind you a little bit of a teeter totter?!

Instead of moving the C.G. forward you might want to move the C.L. back.

Control Surfaces

This can be done by making a CONTROL SURFACE: a part of the plane that catches the air and controls how the plane flies.

If your model has flaps on the tail, bend them down very lightly. You can also roll them on a pencil, or just scrape lightly with your fingernail. The paper should not be sharply creased or folded; it should be a very light change of angle on the wing tip.

As the plane moves through the air, the air pushes against the downward slanting surface and pushes the tail up. It works just the opposite if you angle the surfaces up. (So now you know what to do it your plane is doing a NOSE DIVE!)

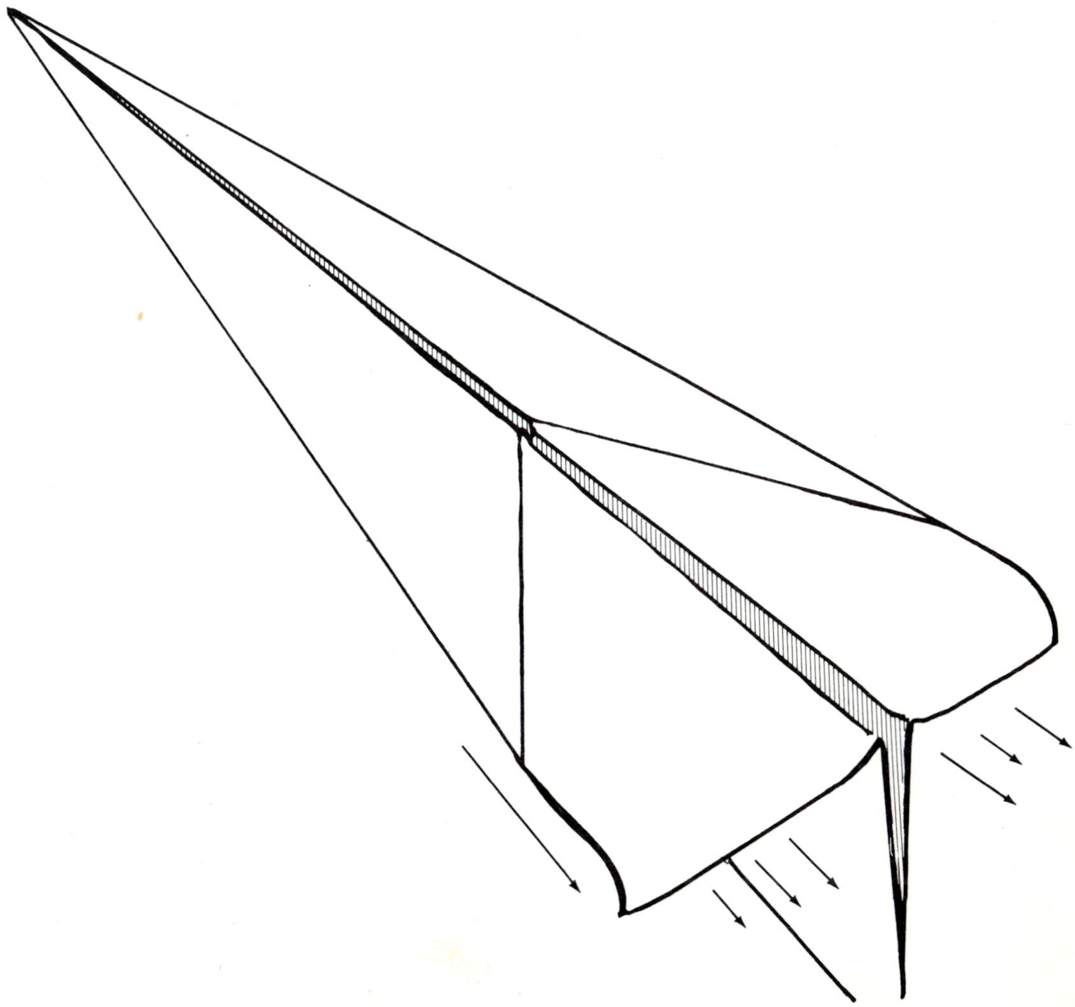

If the nose climbs up and the glider flies in a vertical loop, that is a great stunt. It will happen when your plane has more of its weight on the back end; more lift on the front end and you throw the nose up, kind of hard. You can also make it do a vertical loop going nose down, by launching kind of fast, hard, and nose down with a plane that is balanced with more weight on the nose end and a lot of lift on the back end.

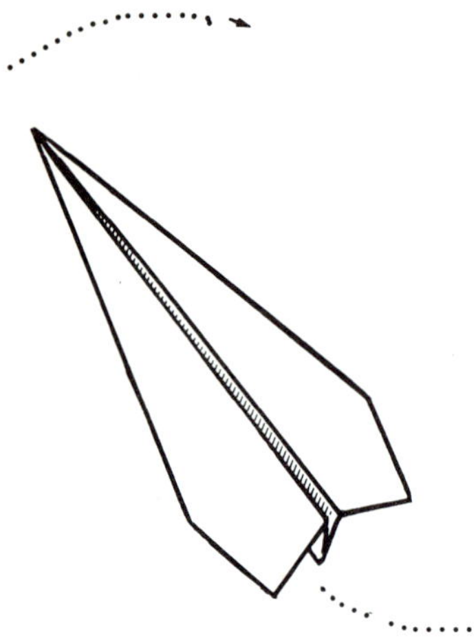

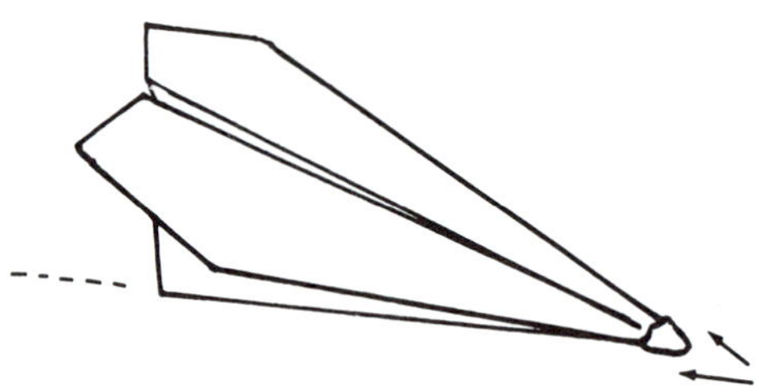

If you don't want it to do a nose up vertical loop, you will need to move the Center of Gravity forward (by adding weight to the nose) or move the Center of Lift back (by bending the tail surface down) depending on what your plane needs.

If your plane doesn't have flap like parts on the tail, just cut or tear, in about a half inch or less from the end of the tail, on the crease, where the wings and tail meet. (See diagram p. 43.)

Keep adjusting the balance little by little until you get a steady, smooth glide. If you add clay or aluminum foil, try to streamline its shape so the air will flow around it easily.

Another way to form the Control Surfaces is to hold your fingers or thumb over the wing tip and against the edge of the table. Then pull the paper through with your other hand. This will curl the tips as much or as little as you want, depending on how hard you press with your fingers and it gives an even curl.

Try bending or curling one side up and one side down. Curl it very gently — just a little bit.

Make several launches alike, one after another. See if your plane will turn toward the side with the curled up surface.

Try curling the tips sharply up and down. This usually makes a plane do snap rolls — where it rolls over on its side again and again.

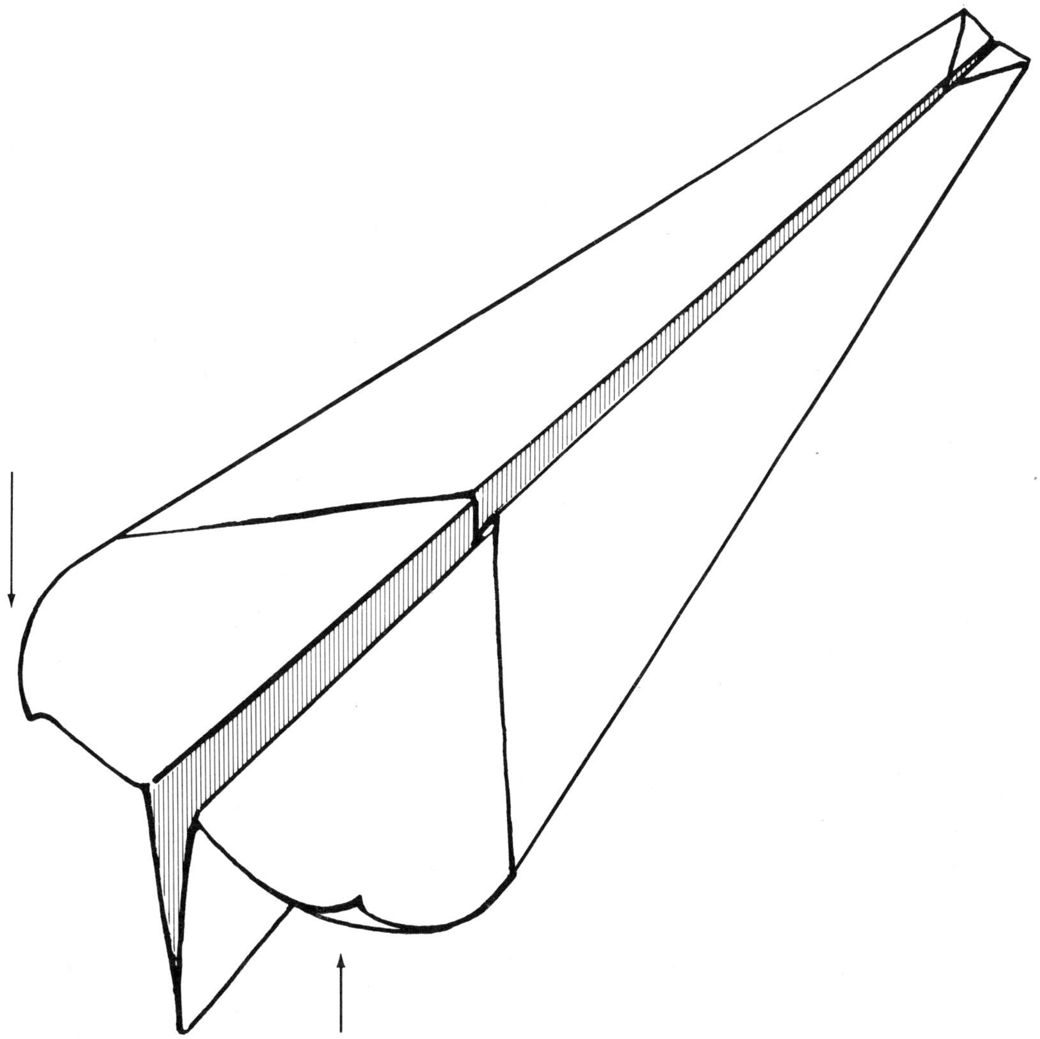

As the plane is flying, moving air rushes against the wings. Some air goes above the wings and some goes below.

If the control surface (called an AILERON) on the right wing is curled up, the air below the wing escapes and won't push that wing up much, while the air above it will push against the curled up AILERON forcing the wing downward.

At the same time, air below the left wing will be caught by the curled down AILERON and it will push that wing up.

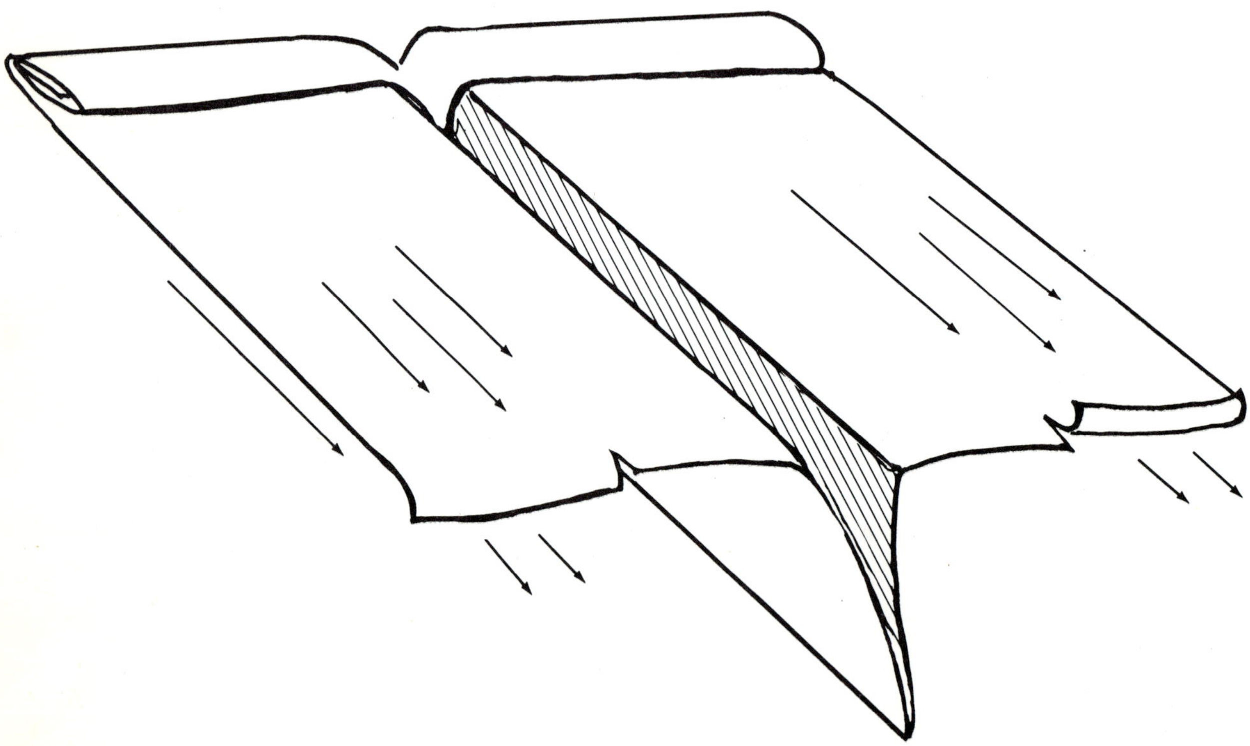

Ailerons, Elevators and Rudders

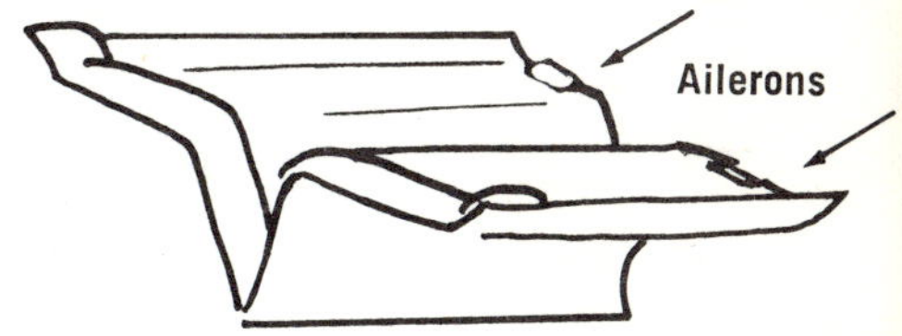

There are three kinds of CONTROL SURFACES on regular planes: AILERONS, ELEVATORS, and the RUDDER.

The Ailerons are located on the trailing edges (back edges) of the wings to make the plane roll over on to the left or right side.

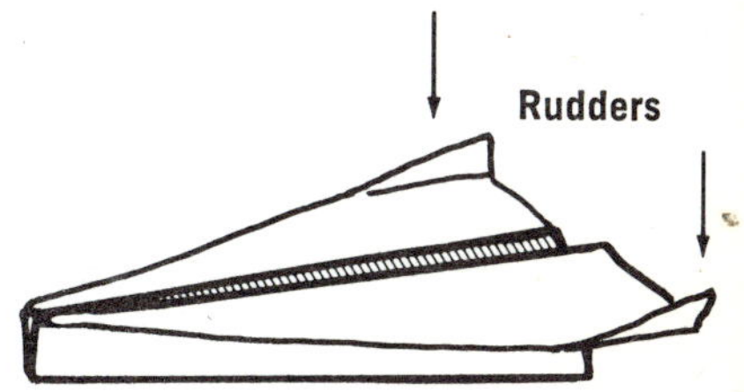

The Elevators are on the horizontal part of the tail; they make the plane nose up, or nose down (dive).

The Rudder is on the vertical tail and is used to make a plane turn toward its left or right.

Since a paper airplane isn't quite like a regular plane, not all of these surfaces will be made in the same places as they would be on a regular plane. However, they are still very effective in controlling the flight of paper gliders.

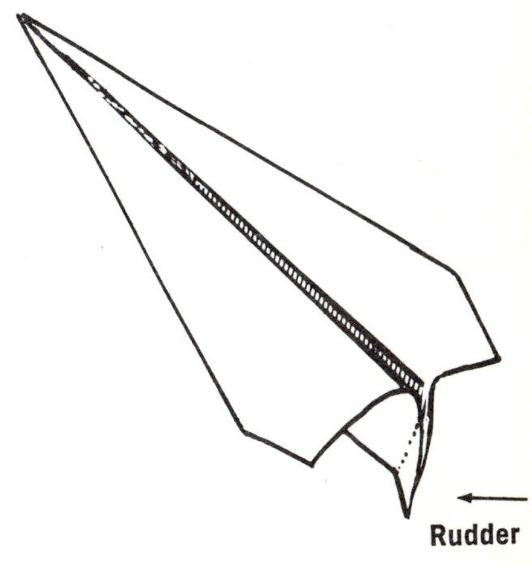

Since the trailing edges of the wings on a paper airplane are usually part of the tail (the horizontal tail), the AILERONS can be used as ELEVATORS at the same time. Many people combine the two words and call them ELEVONS. It is a little harder to maneuver the paper glider with these two things combined, compared to a balsa glider.

The RUDDERS (vertical control surfaces) on some paper airplanes are located along the sides of the wings. On others the vertical control surface is just below the plane wings and horizontal tail. See pp. 43-45 for more about Control Surfaces.

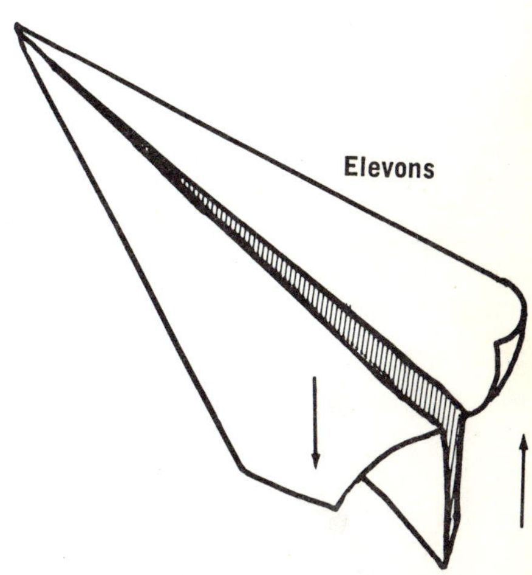

Model Plane
by
Brian Monroney
Open Living School

Hold your paper airplane up at eye level and look at it straight on. Sometimes the wings have been creased more sharply on one side than another. One may flop down or out, while the other stands straight at a right angle. A wing that flops unevenly can act like a Control Surface, making your plane turn one way or another. You can use the wings in just this way to help control the flight of the glider; or you can just check to be sure a lopsided wing doesn't spoil other control surfaces you are trying to use. Either way, it is wise to be aware of the symmetry (or non-symmetry) of your glider's wings.

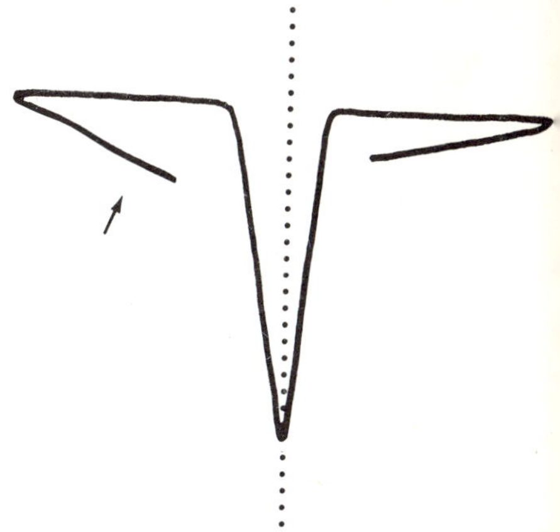

Something else you may want to plan for, is that paper gliders will not hold their shapes for too many flights. You may want to try several different kinds of paper, including magazine pages, or magazine covers as well as more common types of paper.

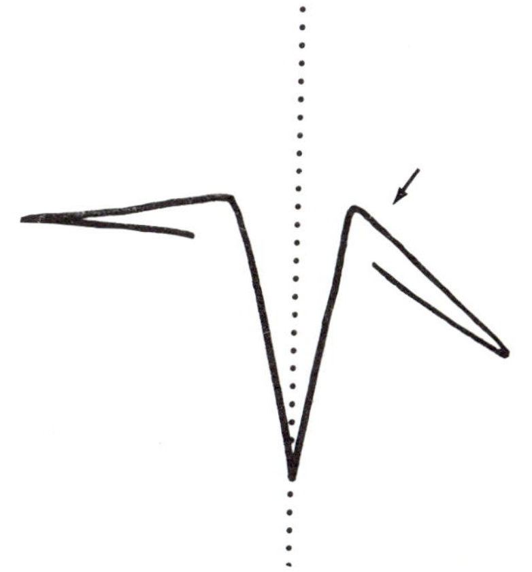

PAPER AIRPLANE PATTERNS . . . .

The following pages have several different patterns for paper airplanes that you may want to try. You will know whether the plane is poroperly balanced, by observing its flight and you can adjust the Center of Gravity by adding paper clips or surfaces on the trailing edges. Test the launch several times. Since each plane launching will be different and will act differently with different ways of launching, you will need to pay special attention to the launch in order to make your plane fly its best.

Model Plane
by
Brian Monroney
Open Living School

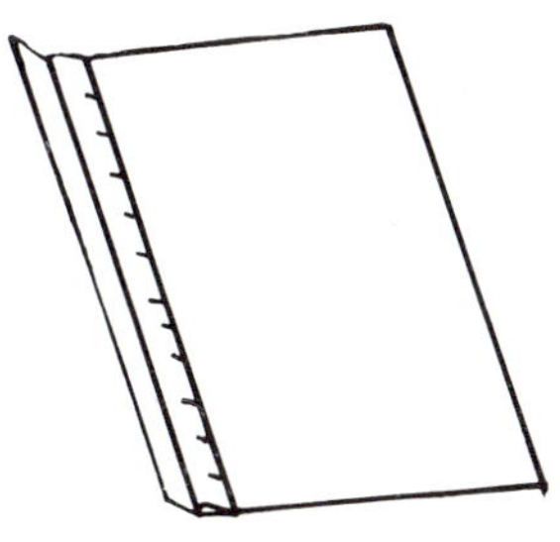

1. Fold the paper about ½ to ¾" over itself, lengthwise, 4 or 5 times. Use a ruler to straighten the fold if you need to.

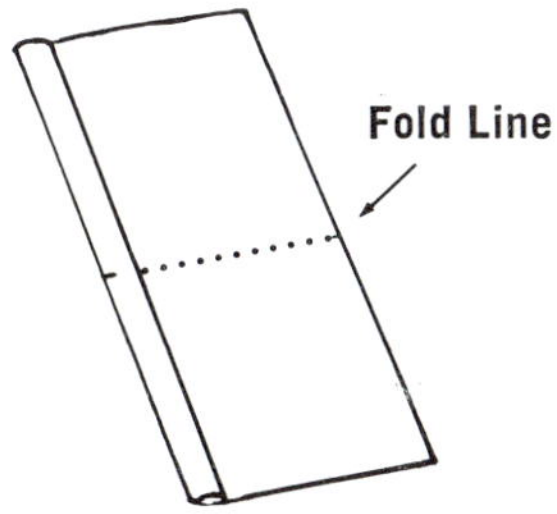

2. Fold it in half crosswise then.

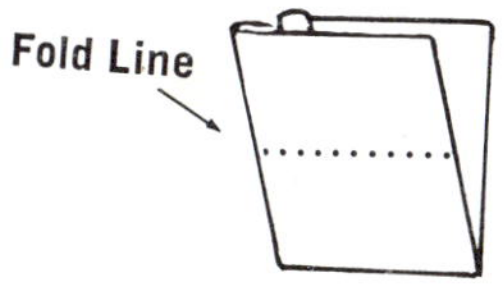

3. Fold the wings back down, in half.

4. Fold up, about ½" in, on the edges of the wings, to make vertical control surfaces. (rudders.)

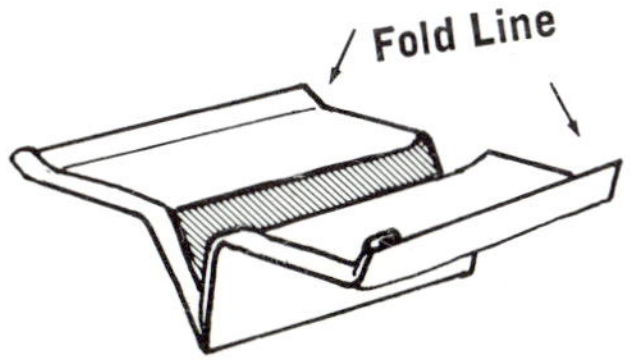

or
Experiment with other ways of folding to make control surfaces. (Fold the rear corners up, or cut slits in the paper to make ailerons, etc.)

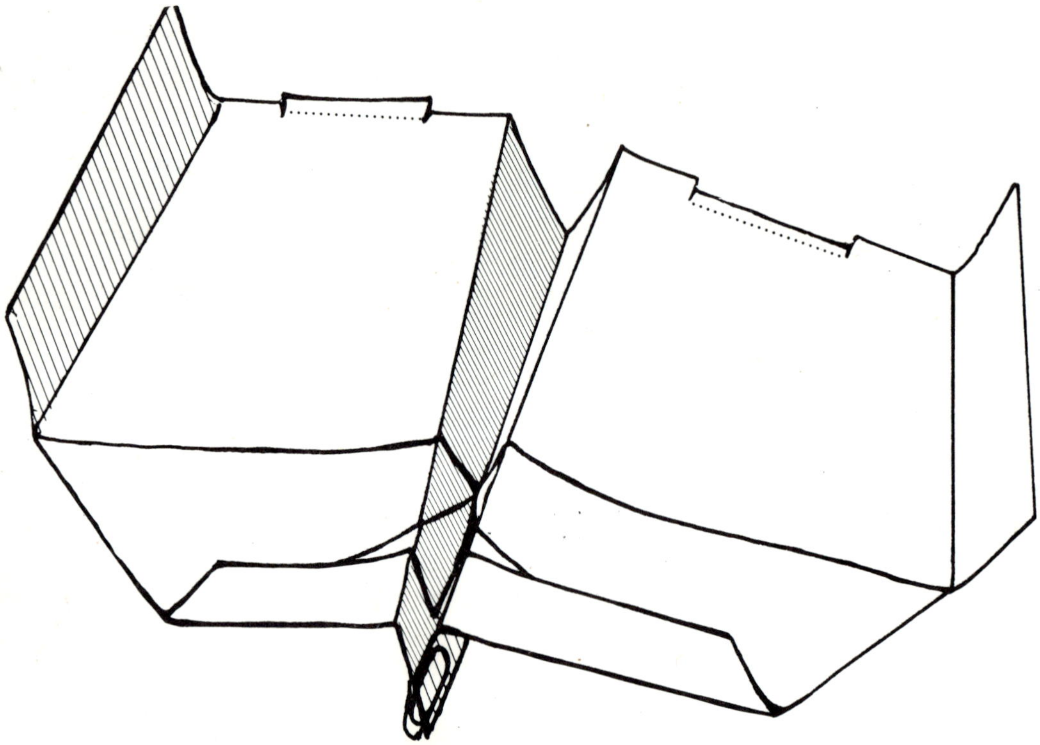

EXPERIMENTAL PLANE
Model by
Danny Byars
Open Living School

Here is one that Danny made. We called Danny the "Experimenter" because he experimented so much with paper airplanes and balsa gliders.

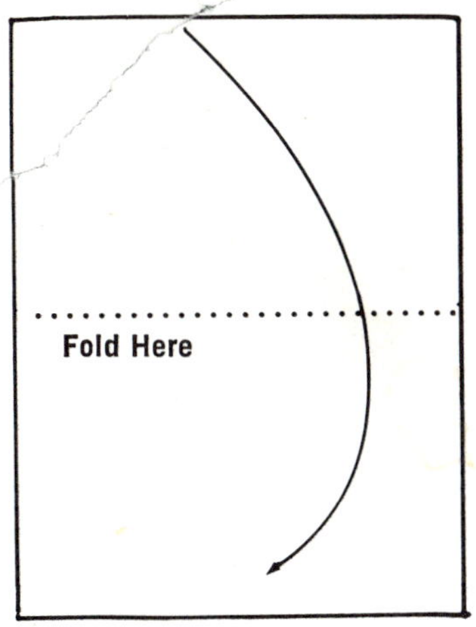

1. Then open it up again.

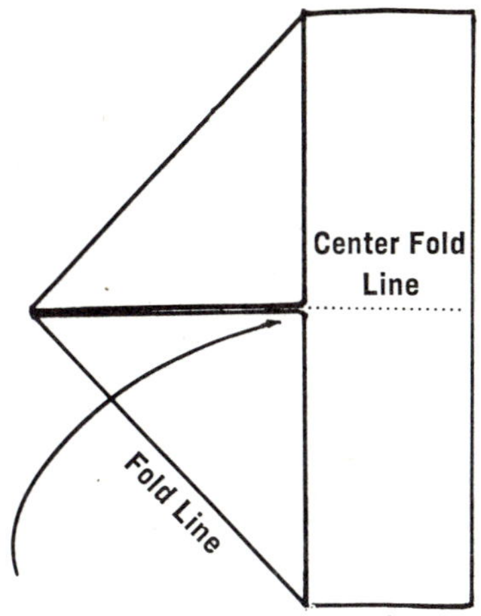

2. Bring corners to the center fold line.

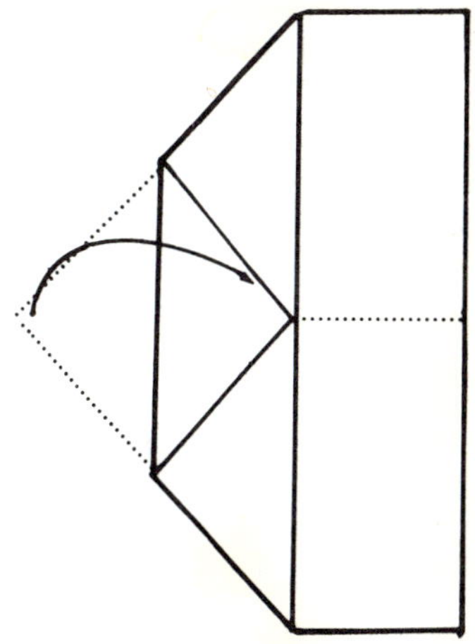

3. Fold outer tip to meet corners on center fold line.

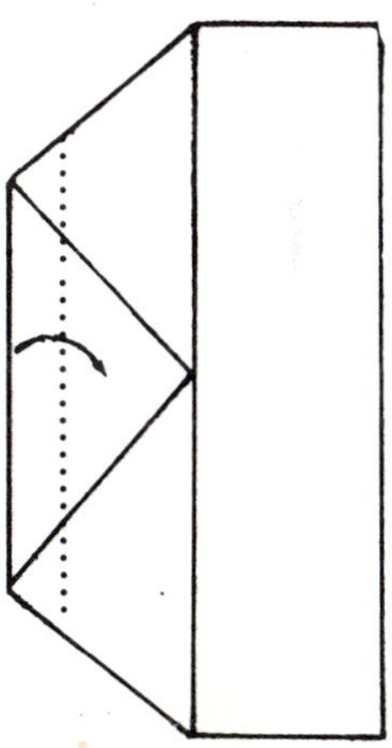

4. Fold about l/3 of the way over.

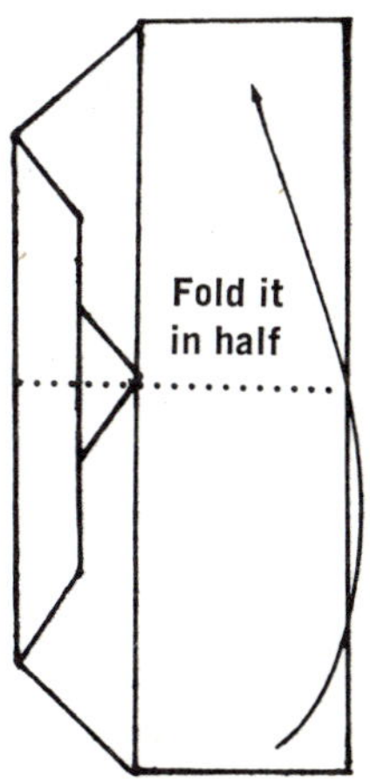

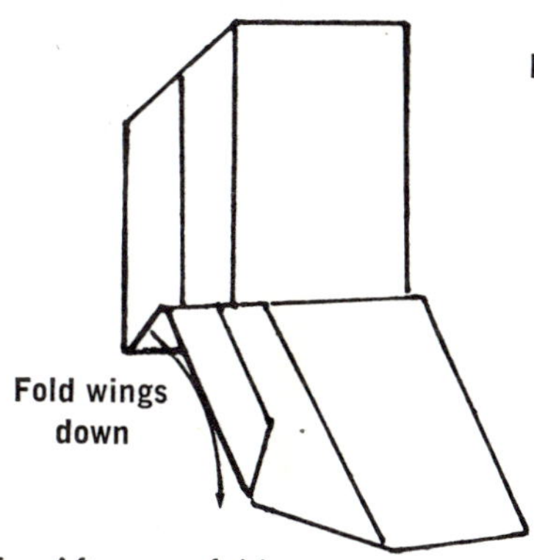

5. After you fold it in the middle, fold each wing down.

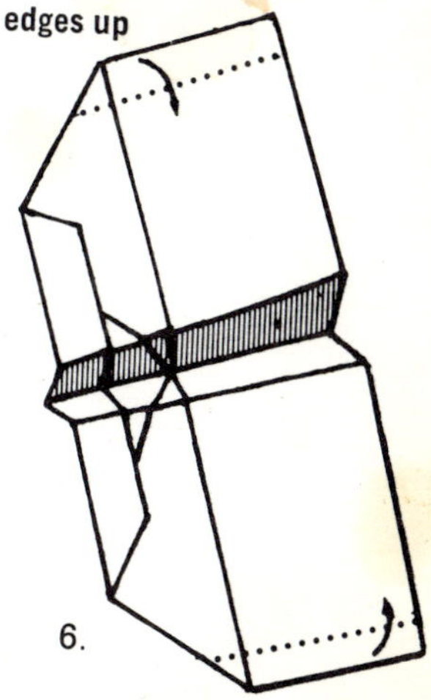

6.

Experimental Plane Model
by
Danny Byars
Open Living School

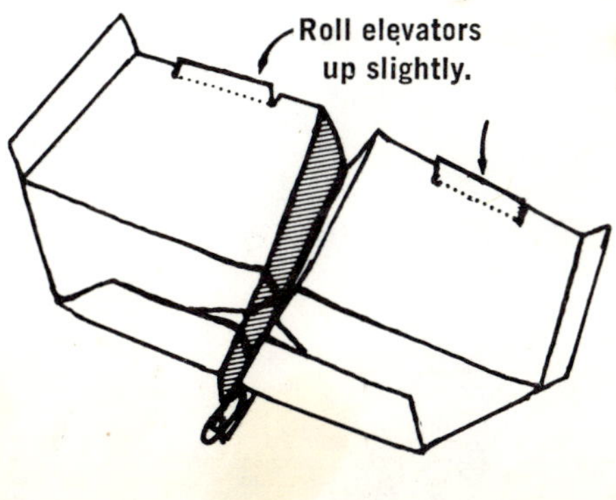

7. Cut in a little less than ½" to make the elevators. Make them about 1½" long. Add a paper clip for nose weight.

"The 5th Grade Model"
by
Barry Kluger
Mountain View Center
University of Colorado

Tiger Tooth Model
by Chuck Hopper
Open Living School

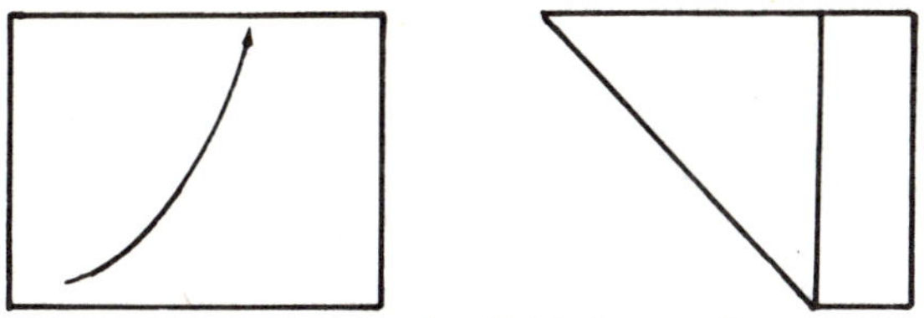

1. Fold diagonally then open it up.

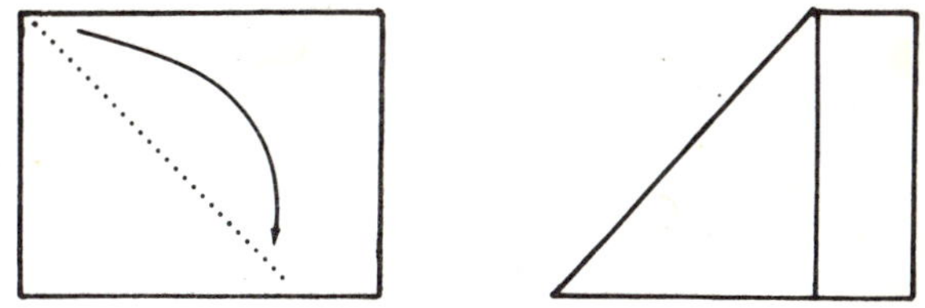

2. Repeat with the opposite corner.

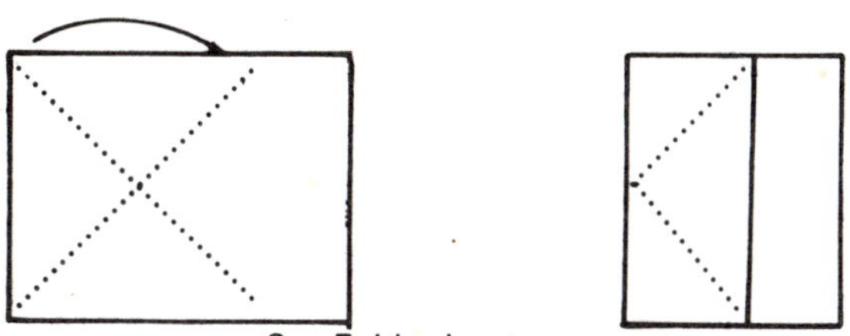

3. Fold edge to corners of diagonal fold lines. Open paper up again.

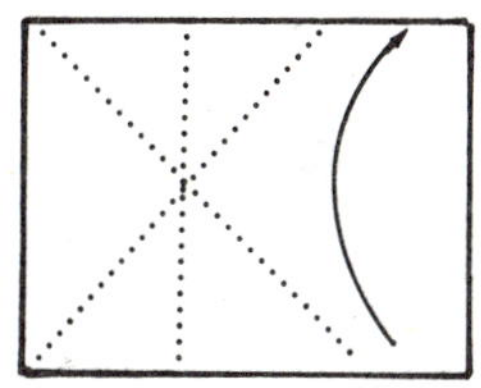

4. Fold in half lengthwise.

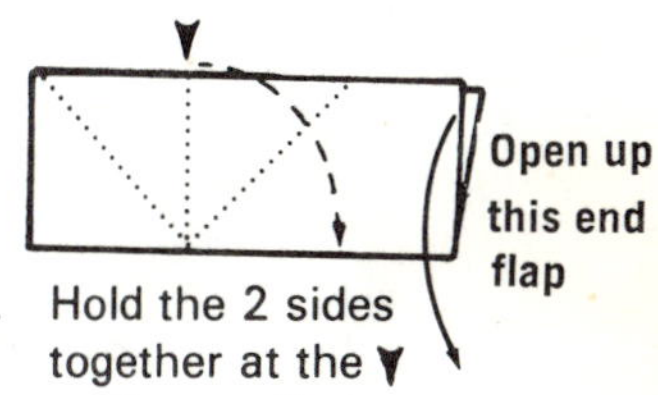

5. Hold the 2 sides together at the ▼ while you flatten the end. See 6.

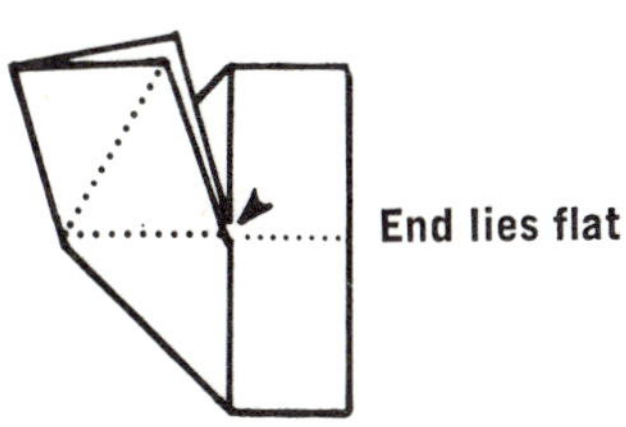

6. ▼ is now on center fold line.

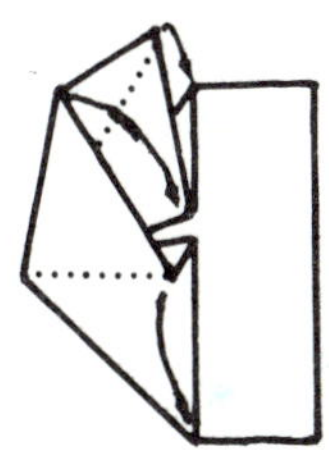

7. Bring upper corners down flat.

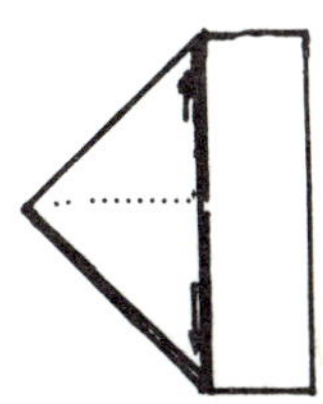

8.

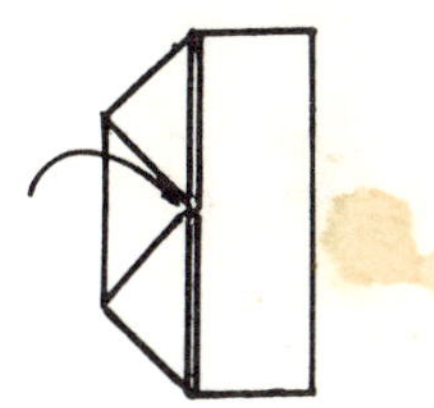

9. Fold the tip over to meet the edges.

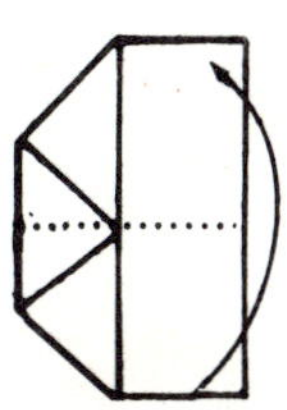

10. Fold the plane in half.

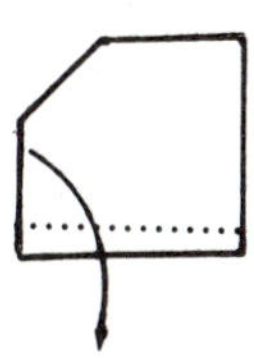

11. Fold the wing flaps down. Make the fold line slant down.

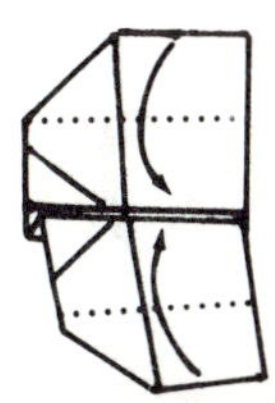

12. Now fold the wing flaps up, in half, on the fold line.

13. Then fold the wing flaps down again, once more in half.

14. Be sure both wings are the same.

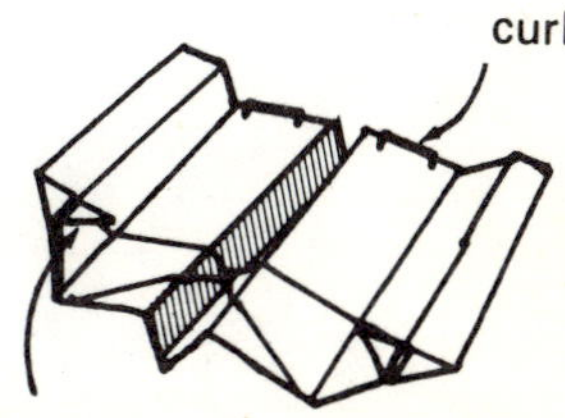

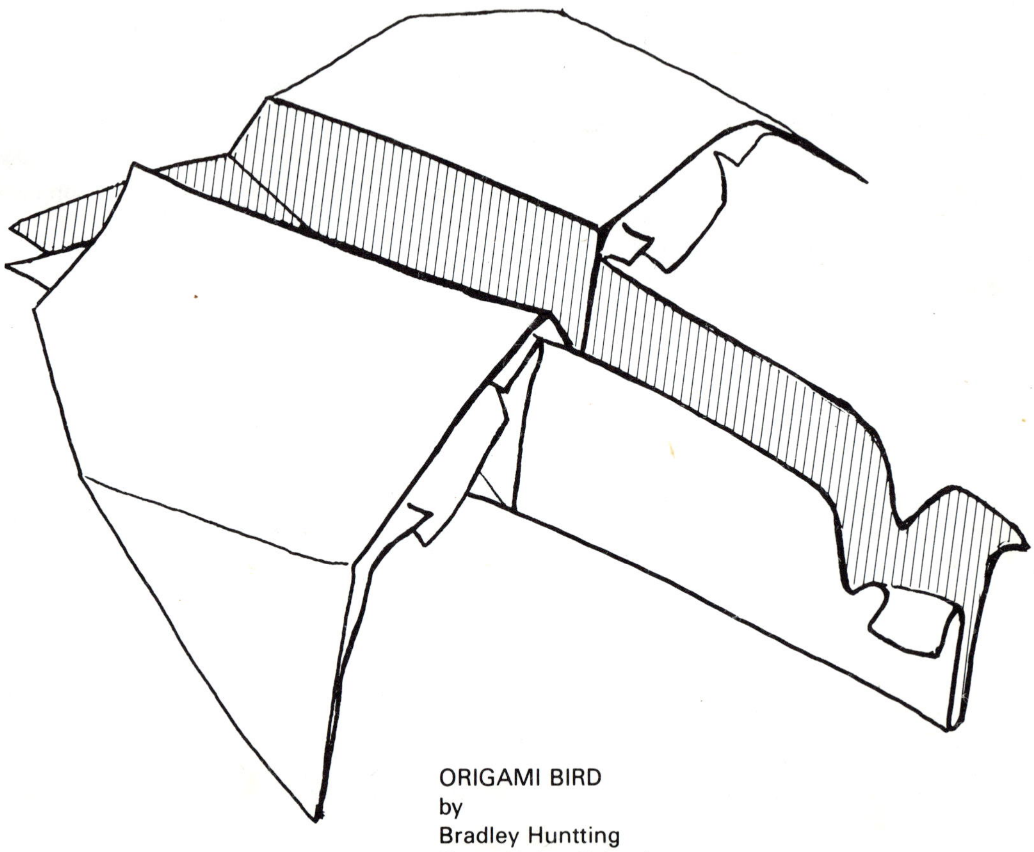

ORIGAMI BIRD
by
Bradley Huntting
Open Living School

Bradley learned to do some Origami (Japanese paper folding) at home. One thing he learned was how to make this bird like paper airplane. It looks like it is hard to do, but that is only because there are so many steps. Believe me, it is worth it to make this one, because it flies so well! The first 6 steps are pretty much the same as Chuck's Tiger Tooth.

Fold diagonally and then open the paper up again.

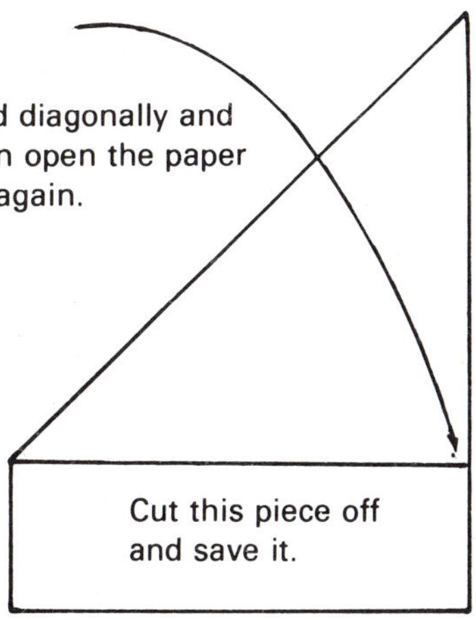

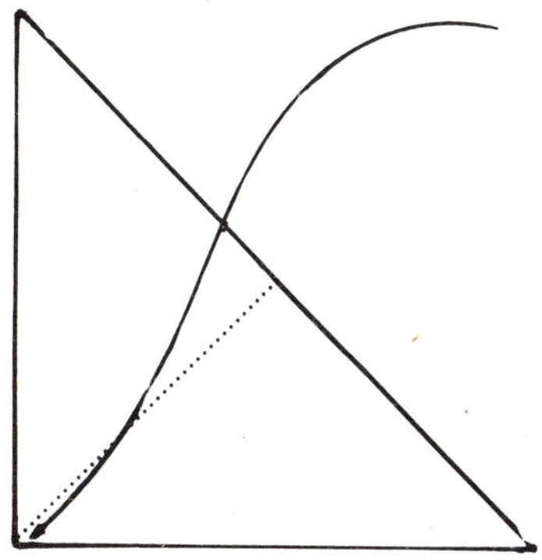

2. Fold again diagonally, the opposite way, and open up again.

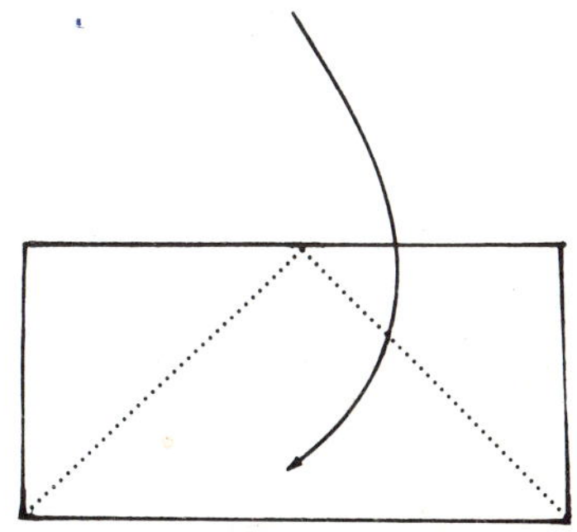

3. Fold in half, and open paper up flat again.

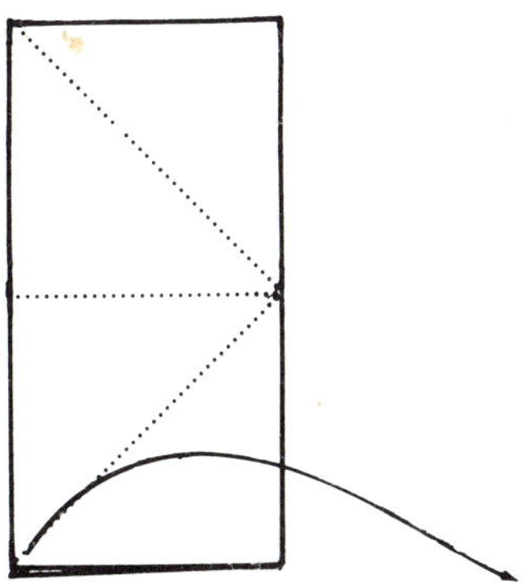

Fold in half vertically; then hold paper together at midpoint, while you open it up flat on the bottom.

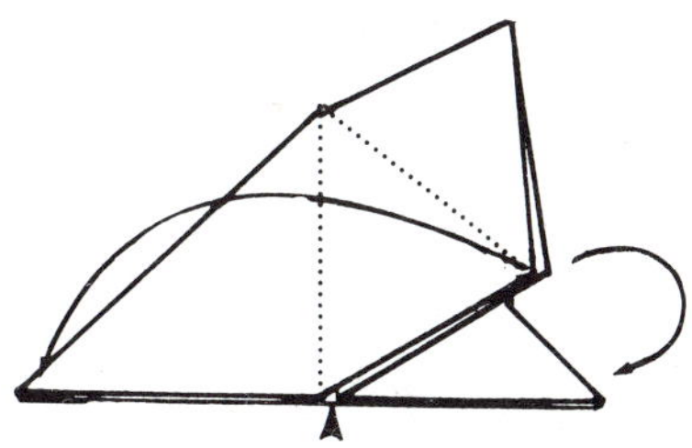

5. Bring the upper corners down flat.

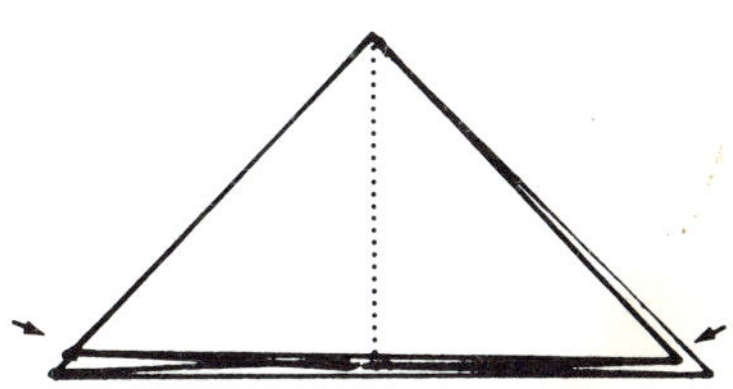

6.

7.

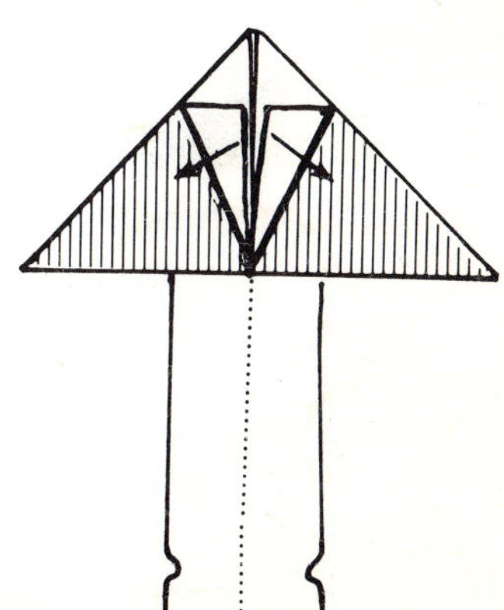

8. Fold lengthwise then cut.

Start about ½'' up. Make only a *very small* cut.

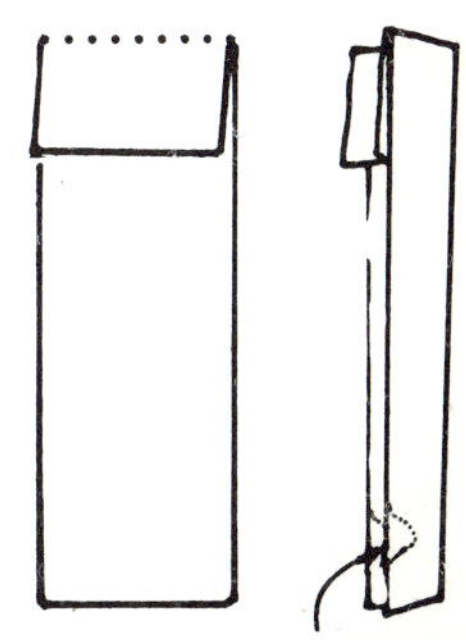

9. Tail piece rests on bottom layer.

10. Fold the lower corners up to the tip of the nose.

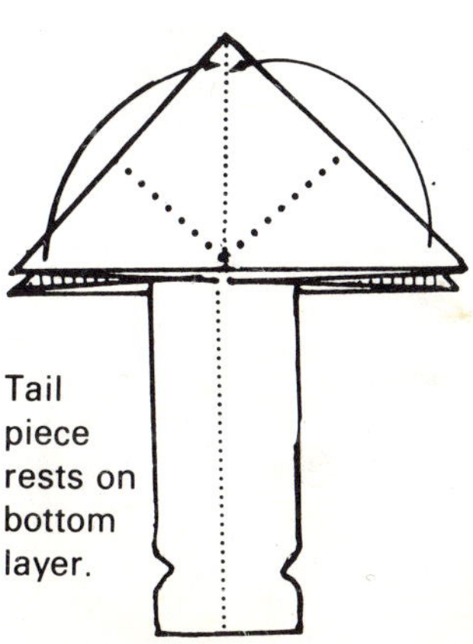

11. Now fold the lower outer edges to the center.

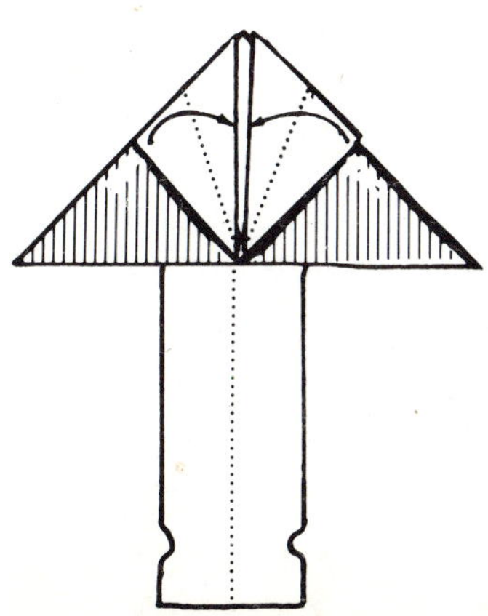

12. Now, unfold step 11, and flatten.

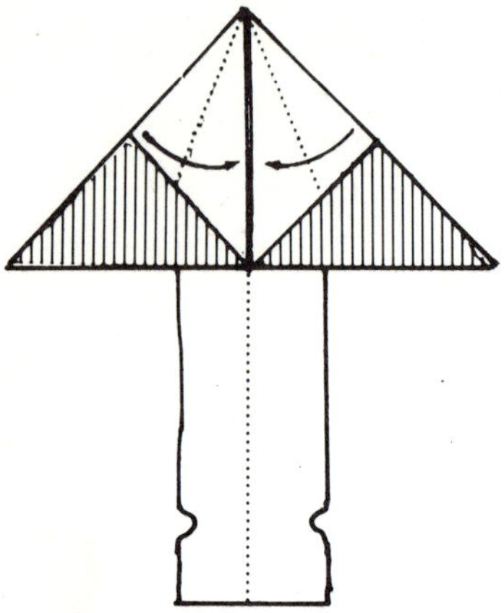

13. Now fold the upper outer edges to the center.

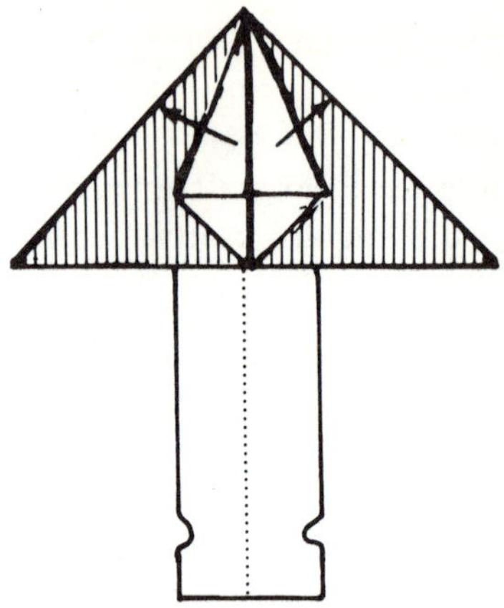

14. And then unfold step 13, and flatten.

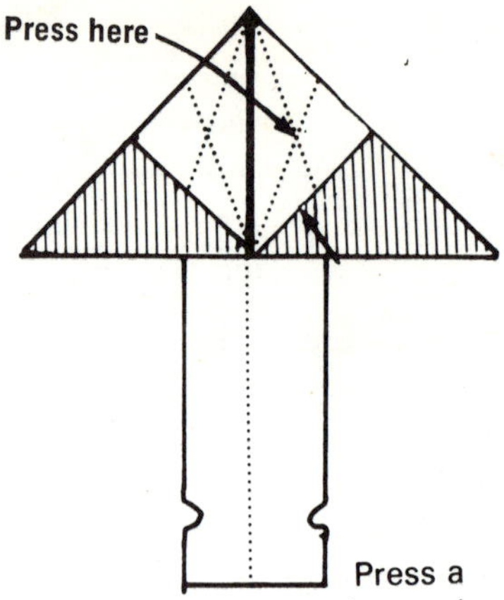

Press a finger down where the 2 folds cross.

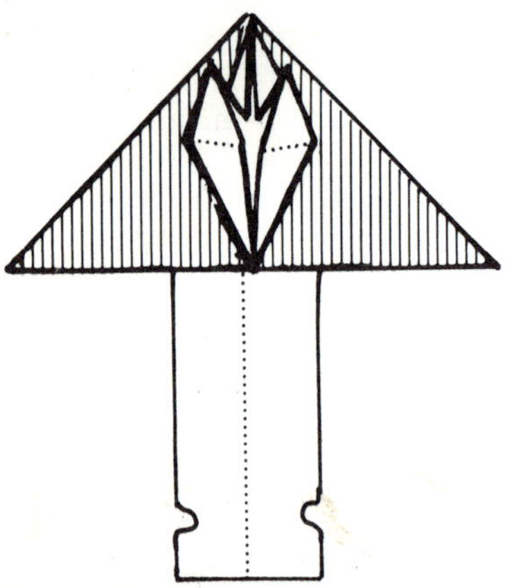

15. At the same time you have a finger pressing down where the folds intersect, using thumb & finger from the other hand, slip under the edges and fold over and let the corners point up. Crease.

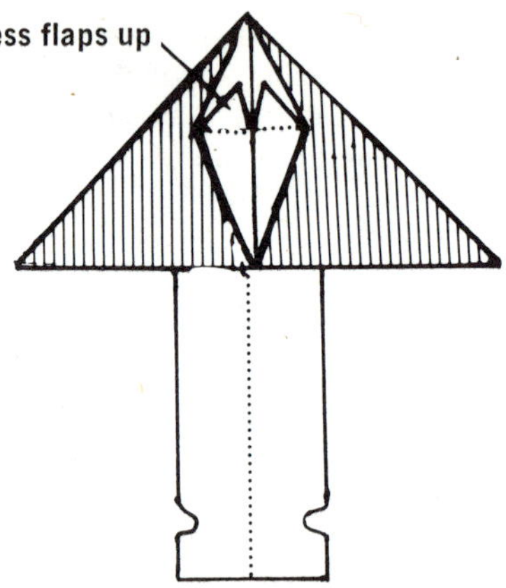

16.

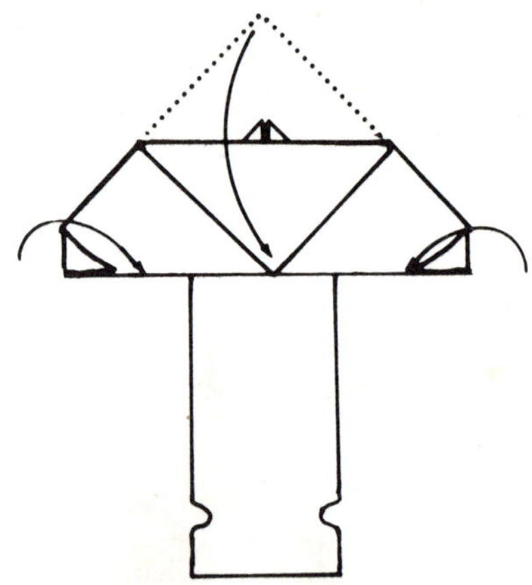

17. Turn to the backside and fold the tips down, and in on the sides.

18. Turn it face up again and fold together in the center.

19. Then fold the wings down, even with the top line of the body.

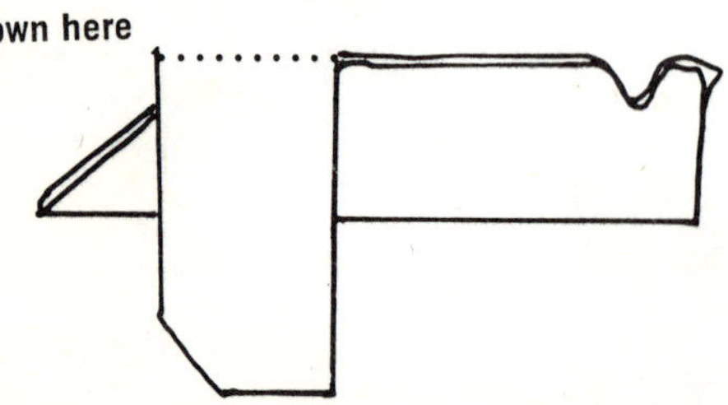

20.

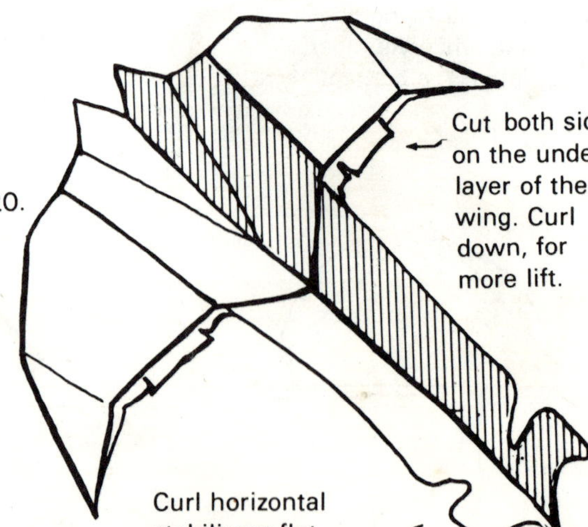

Cut both sides on the underneath layer of the wing. Curl down, for more lift.

Curl horizontal stabilizers flat and slightly up for more "weight" on the tail.

Launch the bird gently, and pretty level as it is very delicately balanced.

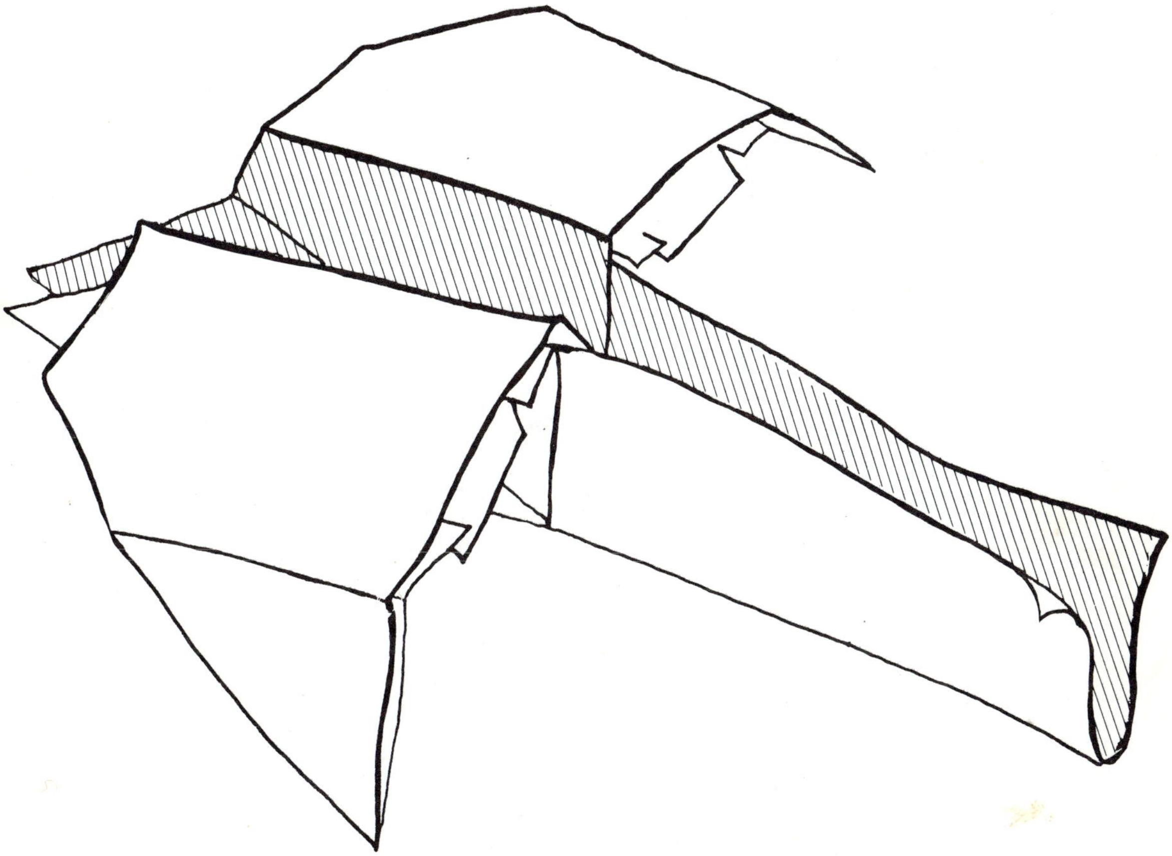

Variation of
the
Origami Bird

Here is a variation of the Origami Bird, which you can make by leaving out step 8, and substituting the following: Fold lengthwise and crease only part of the way down. Leave about ¾" or so uncreased. I found this to be a little more stable in its balance and flight.

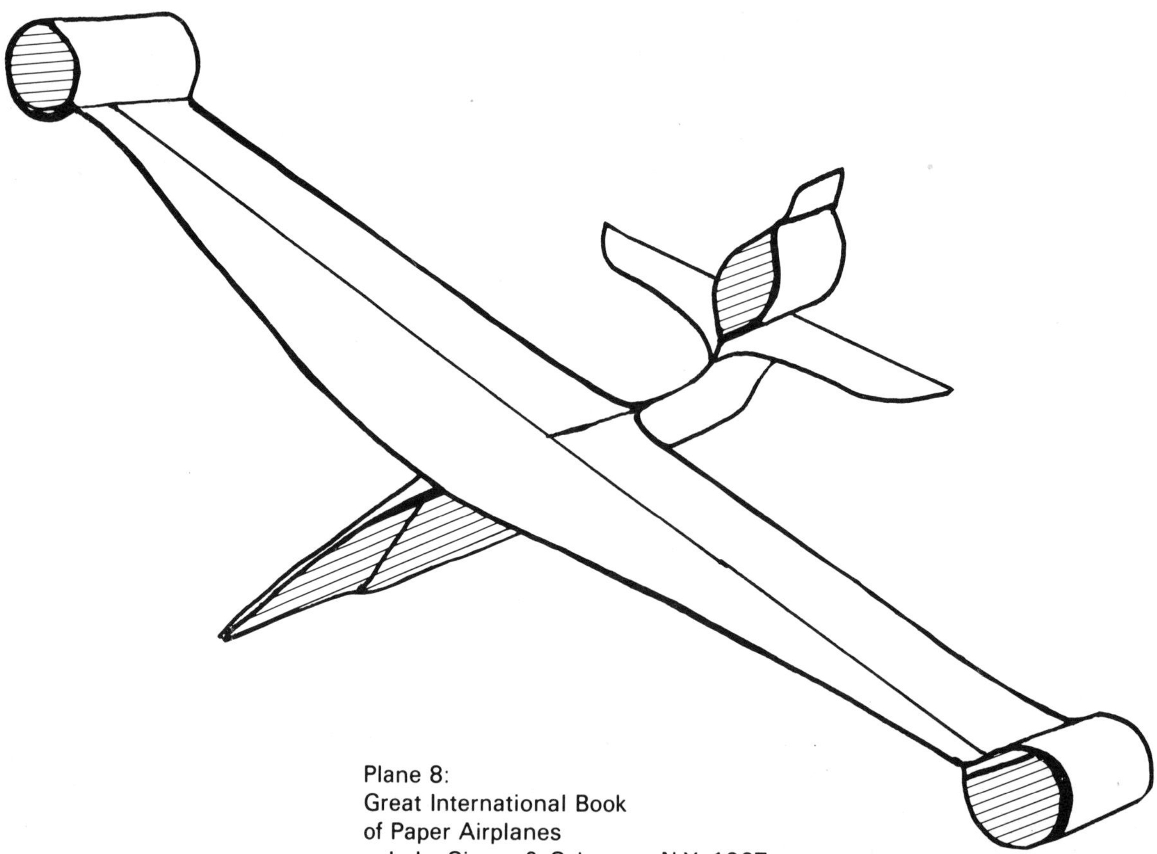

Plane 8:
Great International Book
of Paper Airplanes
pub. by Simon & Schuster, N.Y. 1967
plane by Lewis G. Lowe

Paper airplanes can be made with all kinds of paper, and combined with other things like sticks and straws, too. Try using things like cardboard, newspaper, tagboard, paste sticks, and straws. There are lots of ideas to practice with in books like the *Great International Book of Paper Airplanes* and also in *Stick and Paper Airplanes.* After you have made several planes you will probably start thinking up new ideas completely on your own.

Our study of paper airplanes was only the beginning of a study of flight and only the beginning of what could be done with paper airplanes. The use of different kinds of paper and different sizes, as well as combinations of paper and sticks, provides an opportunity for some very imaginative and creative inventions. These open up a meaningful way for children to study aerodynamic theory and flight. Three dimensional cambered wings can be made by pasting paper wings to paper airfoil shapes and this would enable students to further investigate advanced aerodynamics. The possibilities are enormous and very rewarding!

Some books that will be useful in a study of paper airplanes . . . . . .

Barnaby, Captain Ralph, *How to Make and Fly Paper Airplanes,* New York, Four Winds Press, 1968

Herbert, Don and Ruchlis, Hy *Beginning Science with Mr. Wizard,* New York, Doubleday, 1960

Mander, Dippel & Gossage, *The Great International Book of Paper Airplanes,* New York, Simon and Schuster, 1967

Steidl, Robert H., *Stick and Paper Airplanes,* New York, Abelard Schuman, 1971

Part Two

Building Balsa Gliders

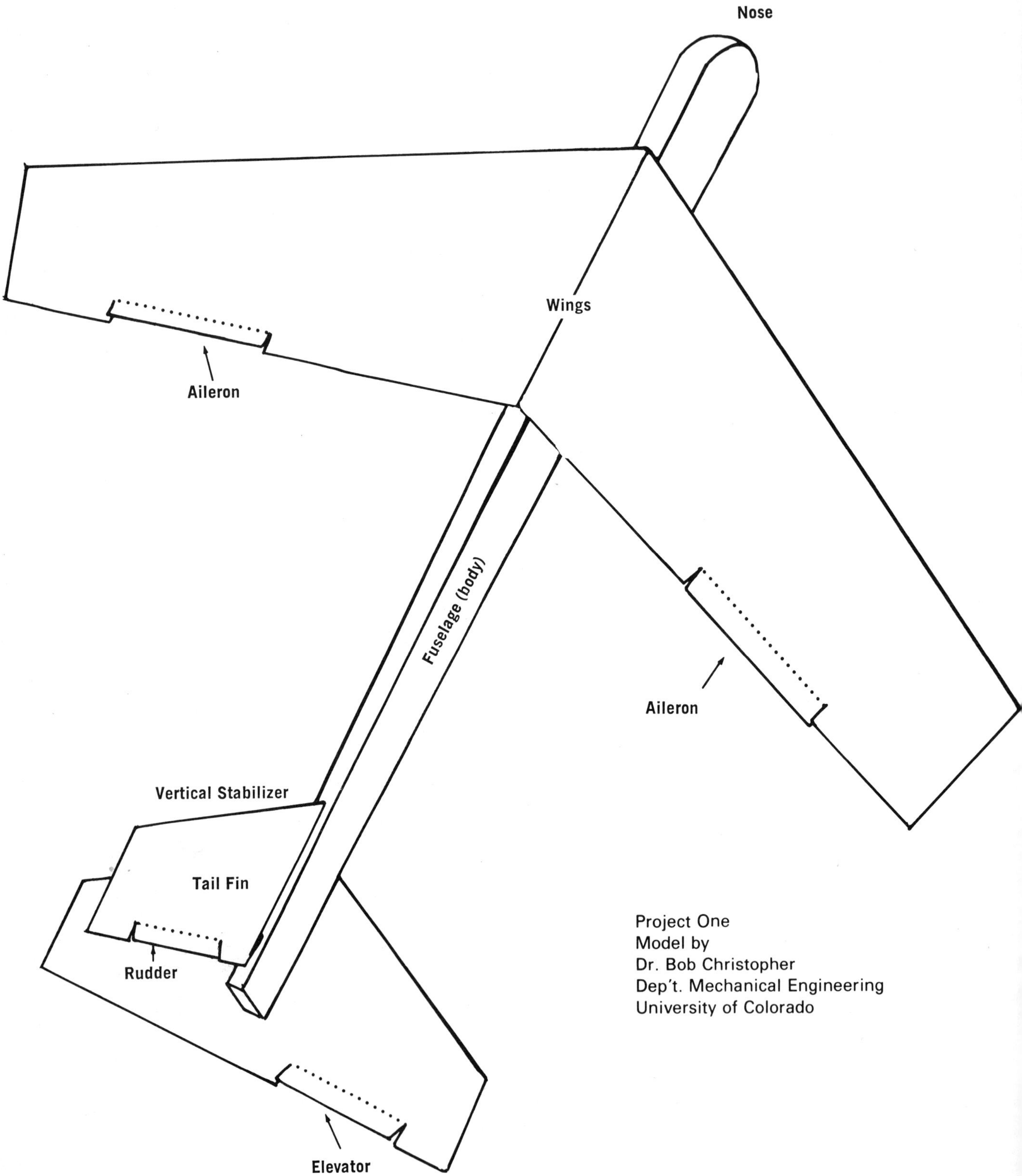

Project One
Model by
Dr. Bob Christopher
Dep't. Mechanical Engineering
University of Colorado

BUILDING BALSA GLIDERS . . . .

When we began our study of balsa gliders we knew very little about them. Books gave us part of our information and we got help from hobby shops and from the *Star Skippers* newsletter (p. 92) as well as the local Soaring Society. But the greatest help we received was from Dr. Bob Christopher in the Department of Mechanical Engineering at the University of Colorado. His personal interest and generous contributions of time and effort in explaining aerodynamics, really gave our glider club the boost it needed to get going in full swing.

If you want to study and build balsa gliders and you want to find out more about them and different models to build, write to the *Star Skippers* newsletter and tell them about your plans. (see p. . .) Also, subscribe to the magazine:

JuniorAmerican Modelers
733 - 15th Street NW
Washington, D.C. 20005

Also, join the:

National Aerospace Education Association
806 - 15th Street NW
Washington, D.C. 20005

It only costs a small amount and you receive a bulletin which contains a lot of useful information.

When you build your balsa glider, you will need the things listed here. Many of them can be found around the house. The rest you can get at the lumber yard, hardware store, hobby shop, or even a discount store.

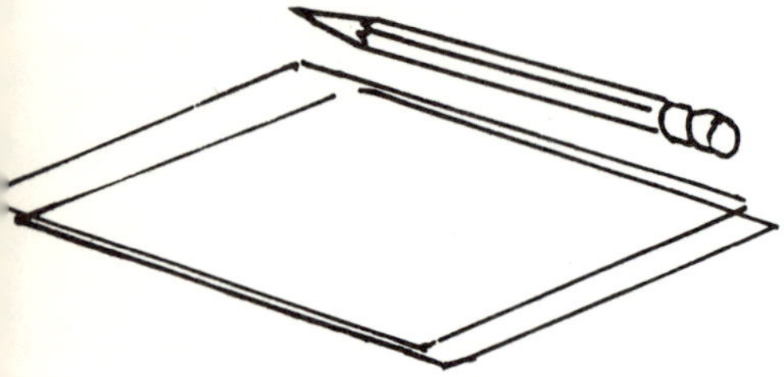

1. Tracing paper and pencil (to trace the pattern out of the book)

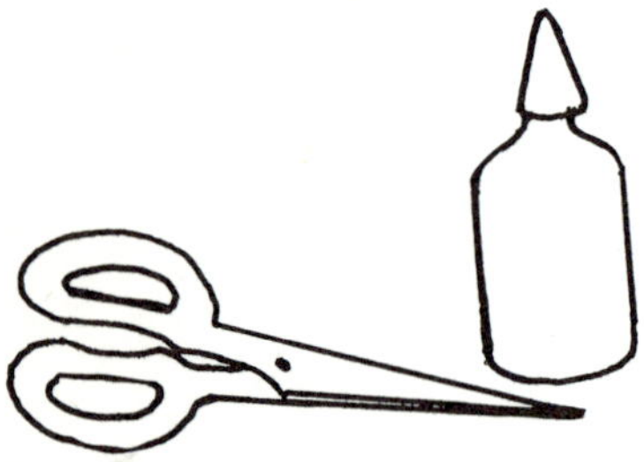

2. Scissors and glue (ELMER'S is fine; fast drying glue is not necessary)

3. Straight pins

4. An exacto knife or single edged razor blade (the exacto is easier to work with, even if you have an inexpensive one)

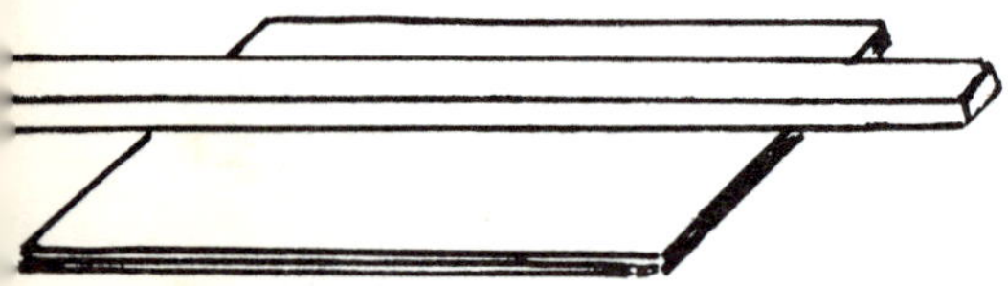

5. Balsa wood: for project one	
	3/16″ x ½″ x 12″ hard balsa
	1/16″ x 2½″ x 8″ soft balsa
	⅛″ x 2½″ x 16″ soft balsa

6. Medium and fine sandpaper and wood blocks to wrap sandpaper around

7. An old magazine or something like it to use while cutting

8. 2 objects, 2 inches high to prop up wings while they dry

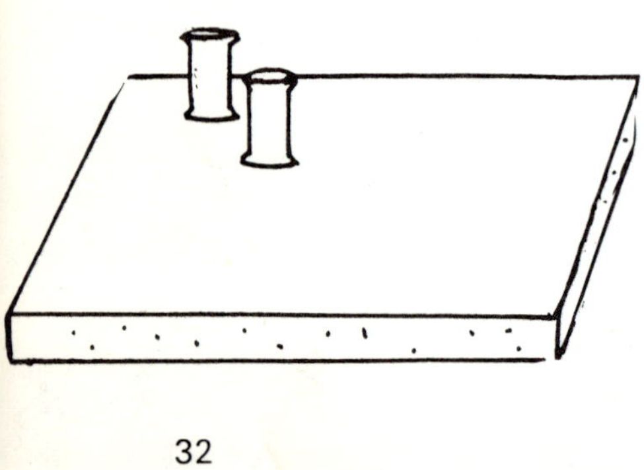

9. Cardboard or old packing-styrofoam to pin parts on while they are drying

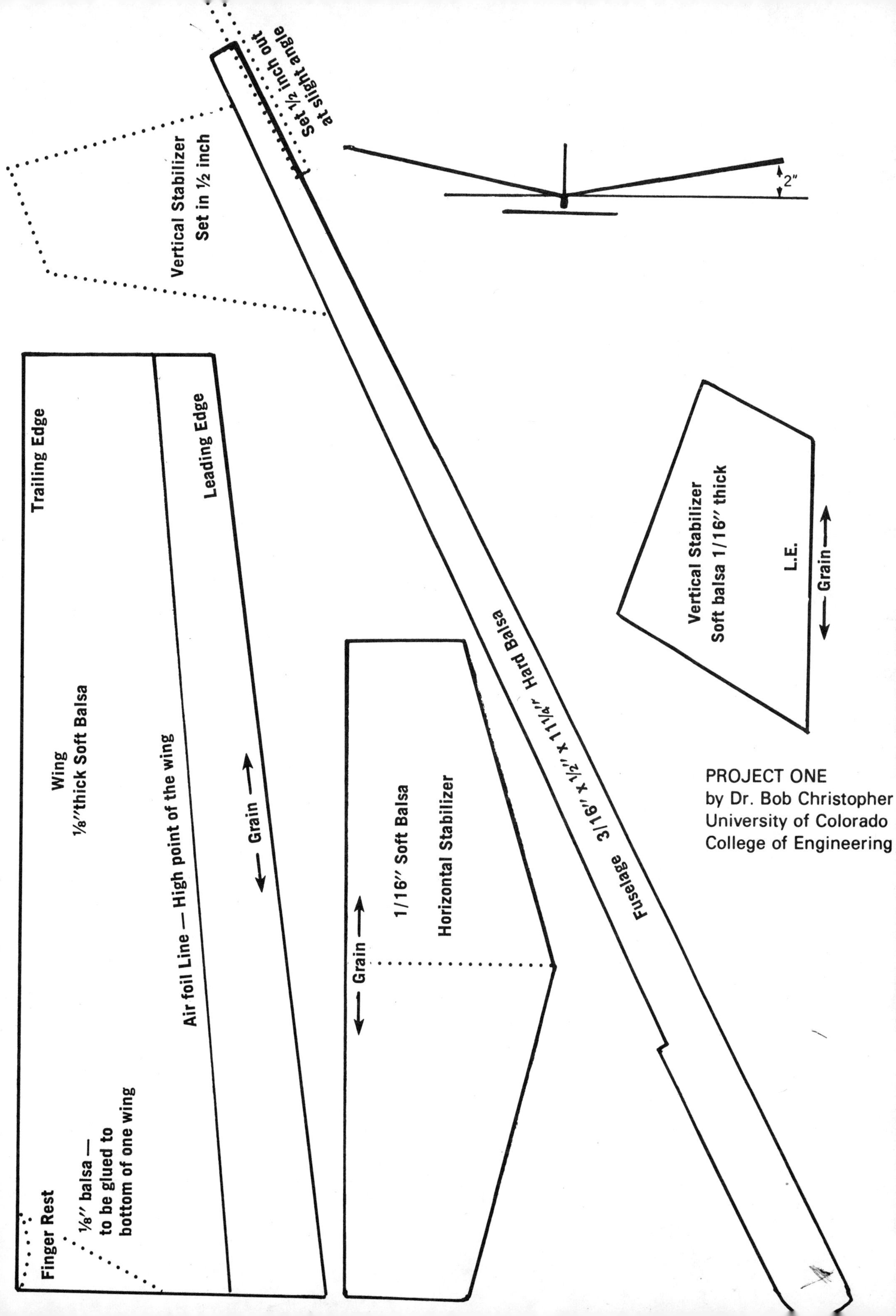

Set 1/2 inch out
at slight angle
Vertical Stabilizer
Set in 1/2 inch
2″
Trailing Edge
Leading Edge
Wing
1/8″ thick Soft Balsa
Air foil Line — High point of the wing
← Grain →
Finger Rest
1/8″ balsa —
to be glued to
bottom of one wing
1/16″ Soft Balsa
Horizontal Stabilizer
← Grain →
Fuselage
3/16″ x 1/2″ x 11 1/4″ Hard Balsa
Vertical Stabilizer
Soft balsa 1/16″ thick
L.E.
← Grain →
PROJECT ONE
by Dr. Bob Christopher
University of Colorado
College of Engineering

BALSA GLIDER PATTERNS . . . .

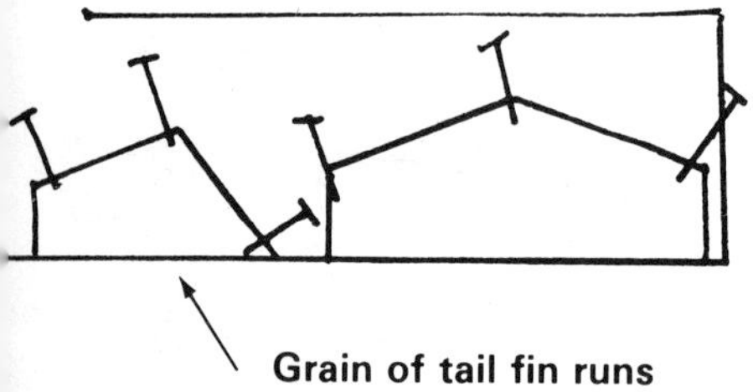

Grain of tail fin runs this way for strength.

Dr. Christopher talked to us about the parts of a glider, about how to make one and how to launch one. He gave us patterns to make them from. This one is the first pattern he gave us and the easiest to build. It is also the pattern many kids used for the gliders they entered in the contests.

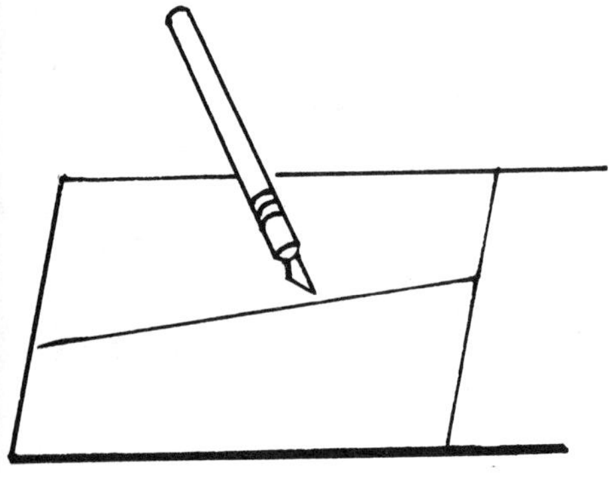

1. Begin by tracing the pattern from the book onto the tracing paper. Cut out the pieces and pin them to the balsa. To keep the pattern flat, slant the pins. You'll have to watch out for them as you cut and take them out as you go. Be sure to put the airfoil line that is on the wing pattern on the balsa wing. This line is to be a guide when you sand the wings

2. Then cut out the balsa pieces. Use the exacto knife or a single edged razor blade. A magazine, or some kind of padding under your work will protect the table.

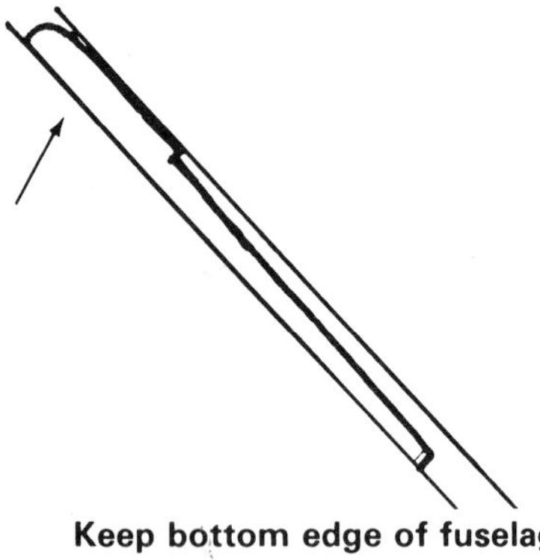

Keep bottom edge of fuselage straight and flat by putting it on the edge of the balsa.

3. Now sand the wings to shape the AIRFOIL. (An airfoil is the shape given to wings so they will be lifted easily by the wind.)

Hold the end of the wing up and look at the edge. It is shaped like a rectangle. Now, sand along the edges of the long sides to make an AIRFOIL.

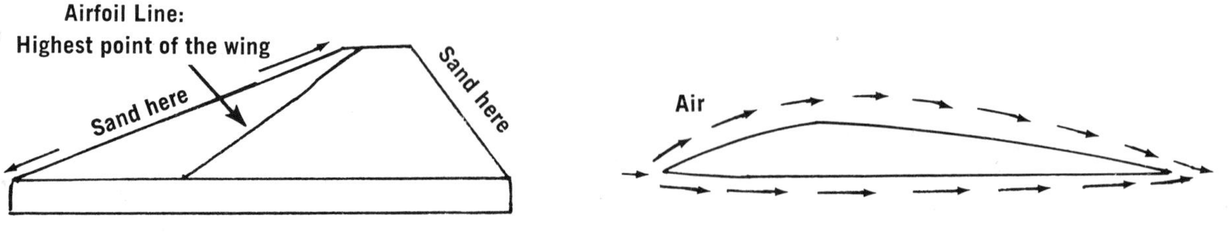

Edge of wing before sanding.

Sand on both sides of the line on the wing. Always keep the line showing. Wrap sandpaper around blocks of wood for a flat even surface. (If you don't use sanding blocks there may be uneven bumps and hollows in your wings.)

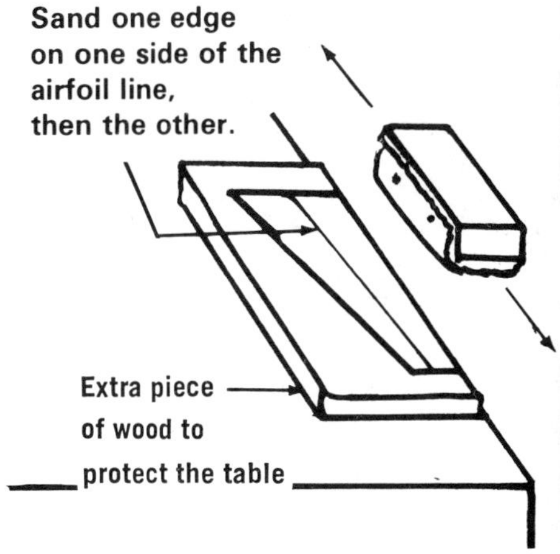

An end-edge of wing after sanding to an airfoil shape. Bottom is flat, top is rounded, tapered down from the airfoil line to the edges.

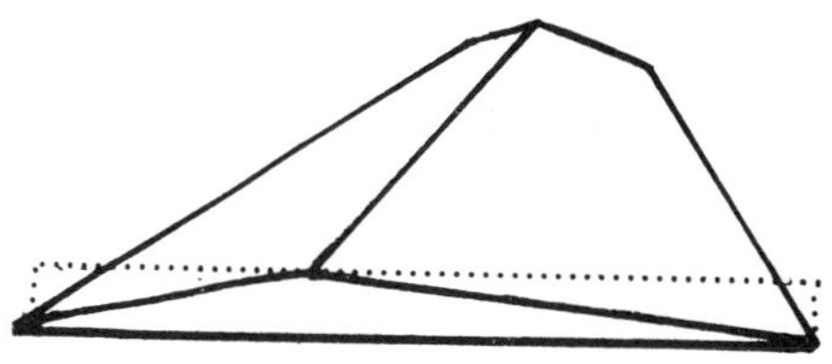

4. With a block 2 inches high, prop up the wing tip. Sand the center end. This will make the wings join smoothly.

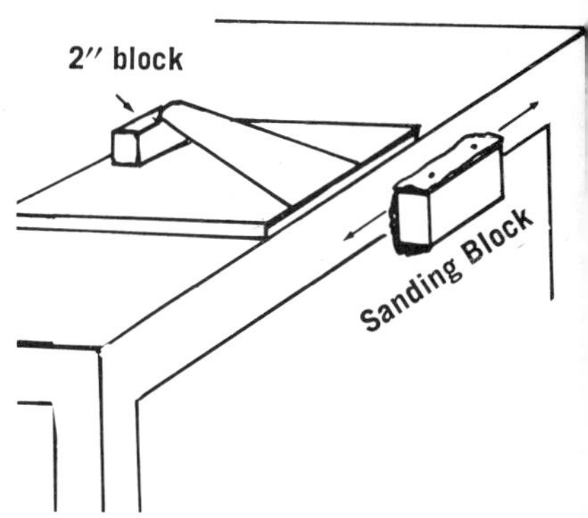

After both wings are together, pin them to the old styrofoam or cardboard. Put objects 2 inches high under the wing tips to keep the angle right until they are dry.

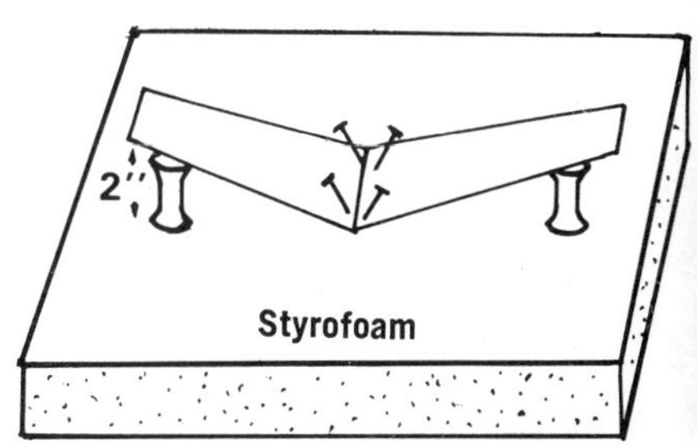

5. Sand the fuselage smooth. Remove the sharp corners and rough edges.

Cut a V-shaped trough for the wings to set into.

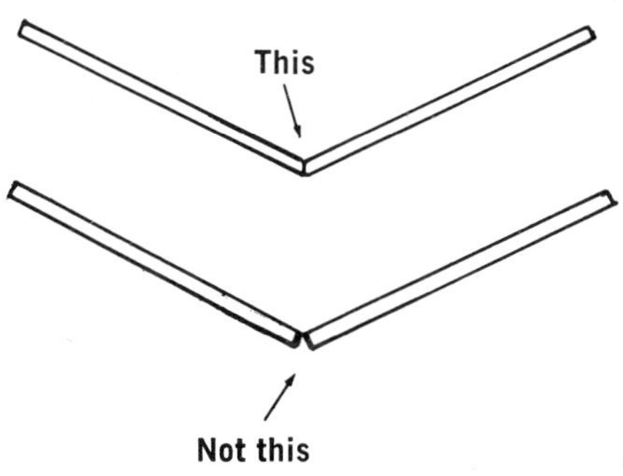

Cut a V-shaped trough for the wings to set into.

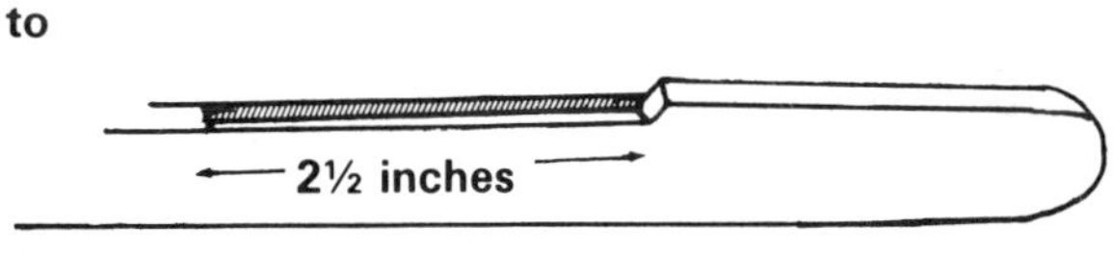

6. Lightly sand the horizontal stabilizer and the tail fin (vertical stabilizer) until they are smooth. Mark the cut for the rudder and cut it out.

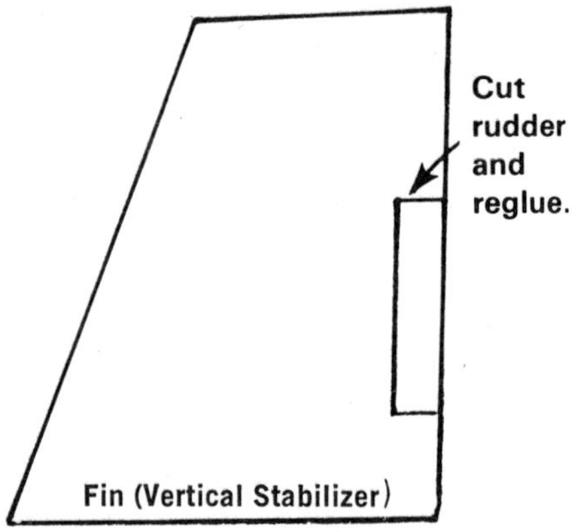

Reglue the rudder, slanting it 1/16 of an inch to the left. (Make it to the right for left-handed launchers.)

7. Sand the last inch of the tail end of the fuselage on the bottom. Sand it to slant upwards at the end. The slant must be very slight. Too much slant upwards will make the glider do loops!

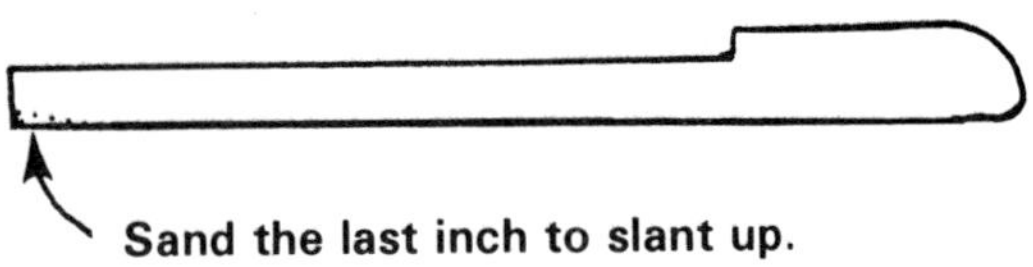

8. Glue the horizontal stabilizer and the tail fin (vertical stabilizer) to the fuselage as shown. The stabilizer will be glued to the bottom and the tail fin to the top.

The fin should stand up straight. The horizontal stabilizer should slant up in back very slightly where the fuselage is sanded.

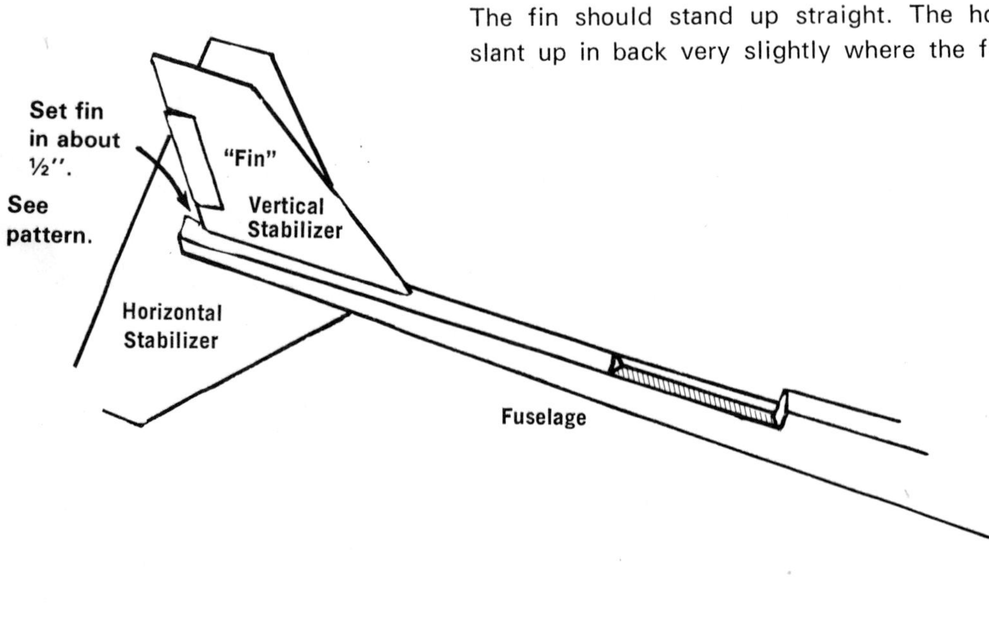

9. Now join the wings (which have already been glued together) to the fuselage, in the V-shaped trough.

10. Glue heavy thread along the Leading Edge of the wings, for extra strength. This is not required, but it will make your glider a little stronger.

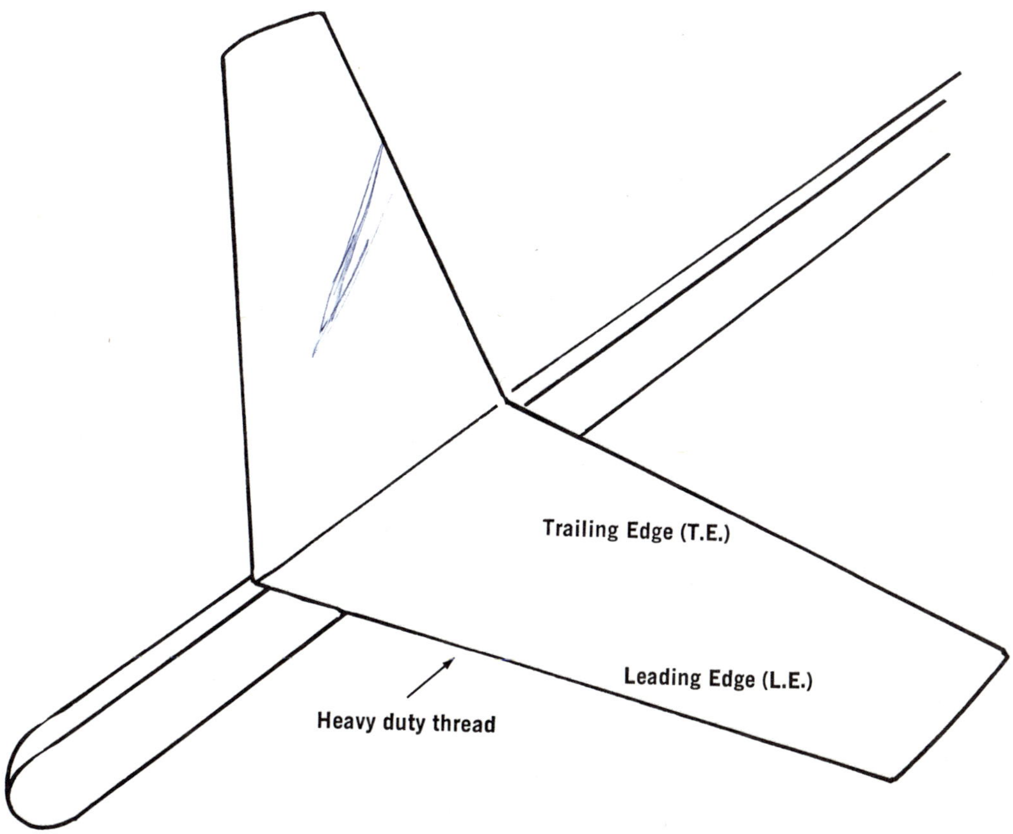

You have now completed the glider!

Put a lump of clay on the nose and try to balance it. Hold the glider with both hands, holding it by the wing tips. There should be just about enough clay to make the nose of the glider almost balance out the weight of the tail. The glider will not really rest horizontally, but rather, will slant down slightly from nose to tail.

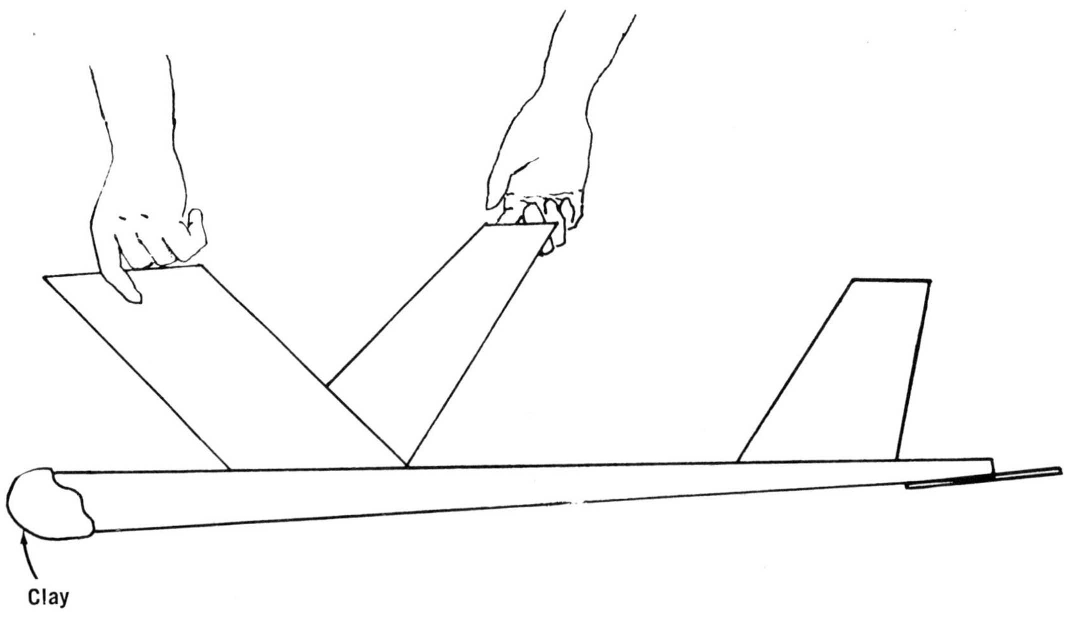

There are some more difficult patterns for other balsa gliders on the following pages. They are fun to build because they fly well. It's worth it to make them, even though they are more difficult.

When you have had experience building several gliders, you will probably want to try to make this adjustment for even better flight. Dr. Christopher explained this to me. To do it you will need a protractor, thin string and a weight for the end of the string. The idea is for the horizontal stabilizer to slant up, about two degrees more than the wings.

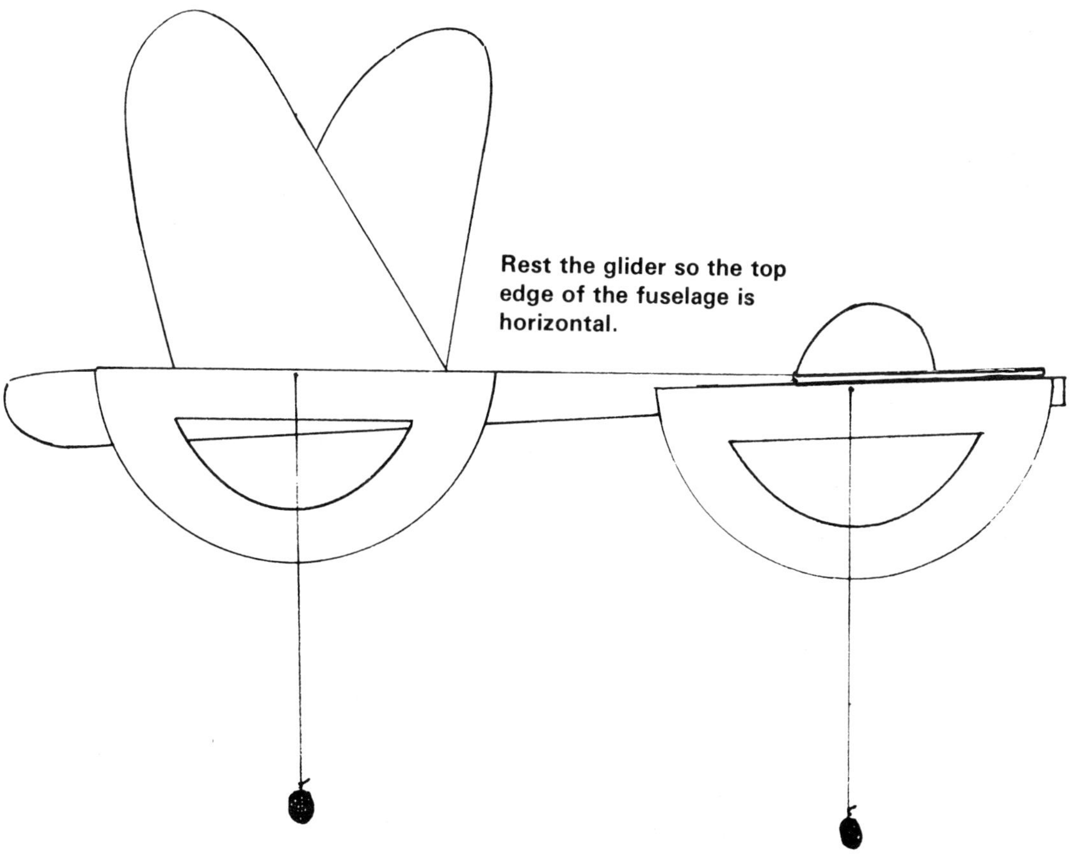

The wings might not rest on the fuselage exactly flat. Perhaps there is an angle of one degree, so the string might hang down right at the line marked 89 degrees. In that case the horizontal stabilizer should slant enough that the protractor will be marked at 91 degrees.

HEDGEHOPPER
by
Warren Kurth

pattern given to us
by Dr. Bob Christopher
University of Colorado
College of Engineering

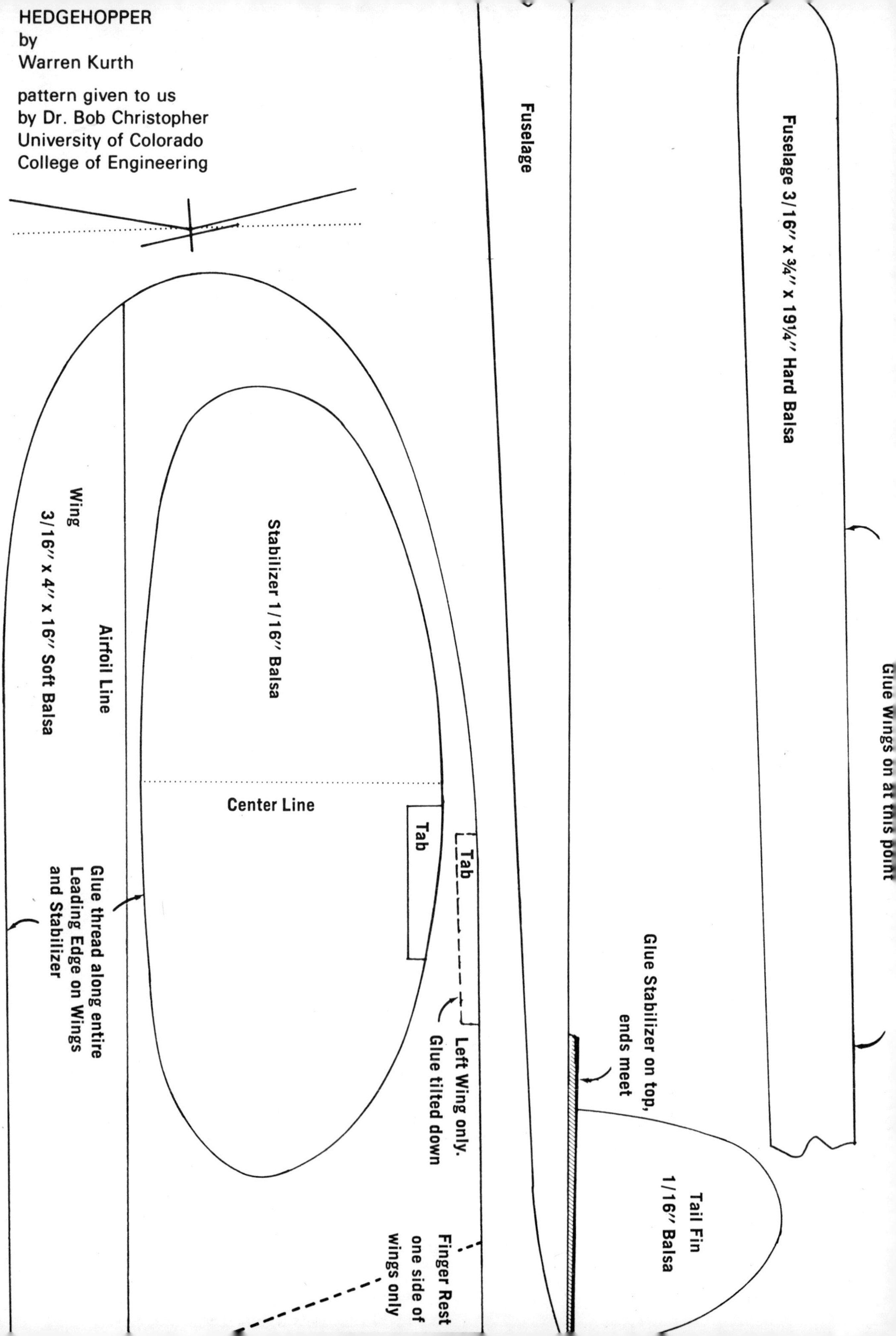

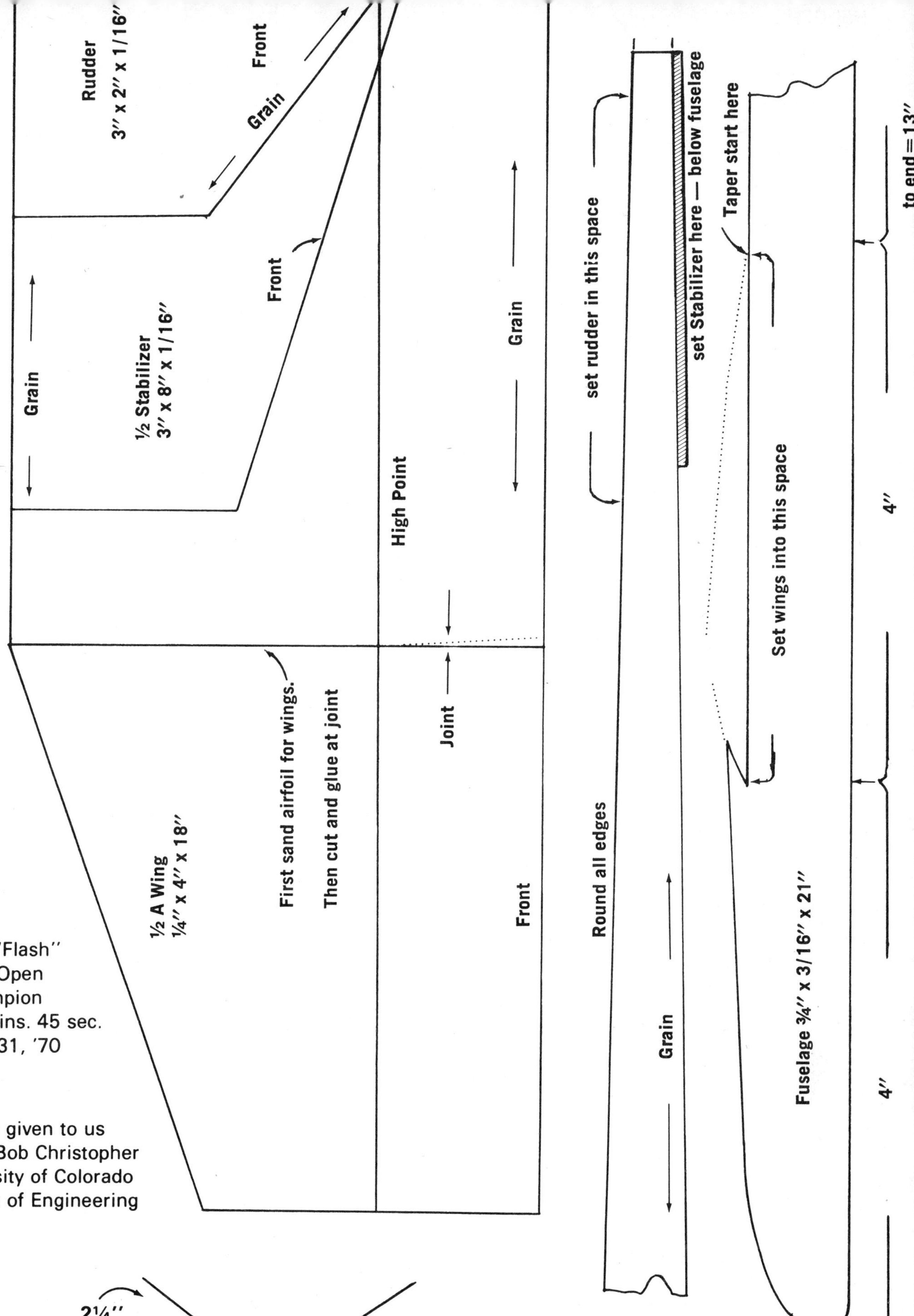

The "Flash"
U.S. Open
Champion
10 mins. 45 sec.
May 31, '70

Pattern given to us
by Dr. Bob Christopher
University of Colorado
College of Engineering

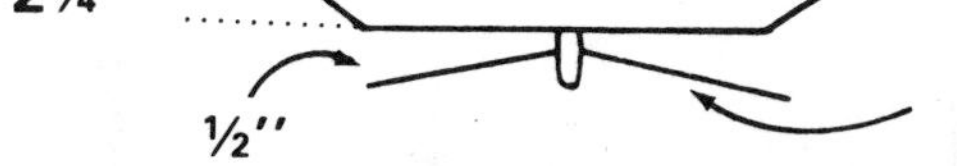

Note!
Stabilizer is down.
However this worked pretty well being tilted more upward too.

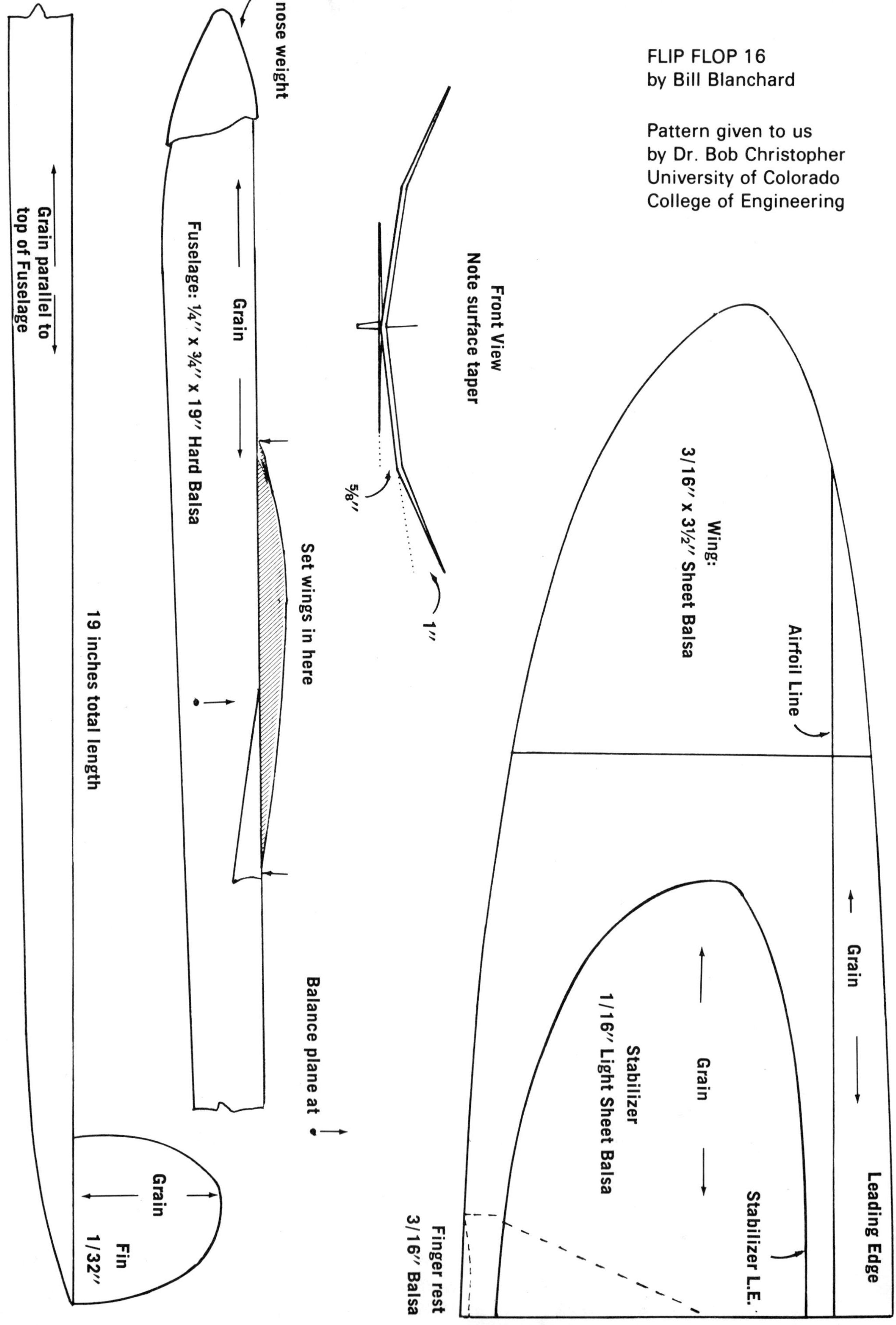
FLIP FLOP 16
by Bill Blanchard
Pattern given to us
by Dr. Bob Christopher
University of Colorado
College of Engineering
lay nose weight
Grain parallel to top of Fuselage
Grain
Fuselage: 1/4″ x 3/4″ x 19″ Hard Balsa
19 inches total length
Set wings in here
Balance plane at
Front View
Note surface taper
5/8″
1″
Grain
Fin
1/32″
Wing:
3/16″ x 3½″ Sheet Balsa
Airfoil Line
Grain
Leading Edge
Stabilizer
1/16″ Light Sheet Balsa
Grain
Stabilizer L.E.
Finger rest
3/16″ Balsa

Control Surfaces are the movable parts of the tail, stabilizers, and wings. By using them, you can control the plane or glider and make it turn, land, and do stunts.
Control Surfaces can also be put on paper airplaines to make them do just about the same things.

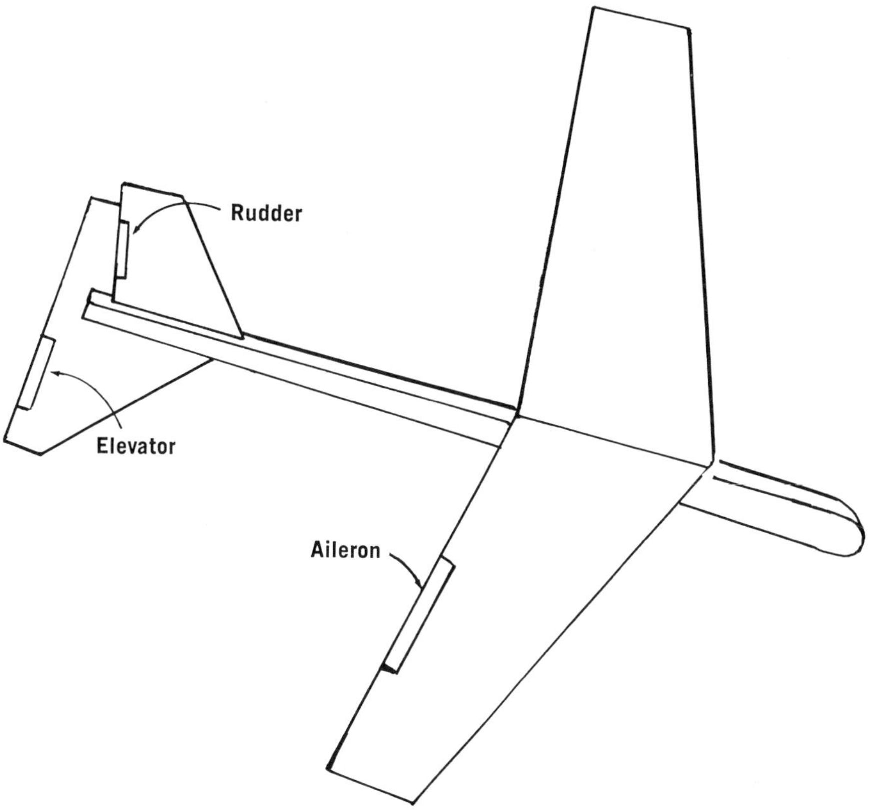

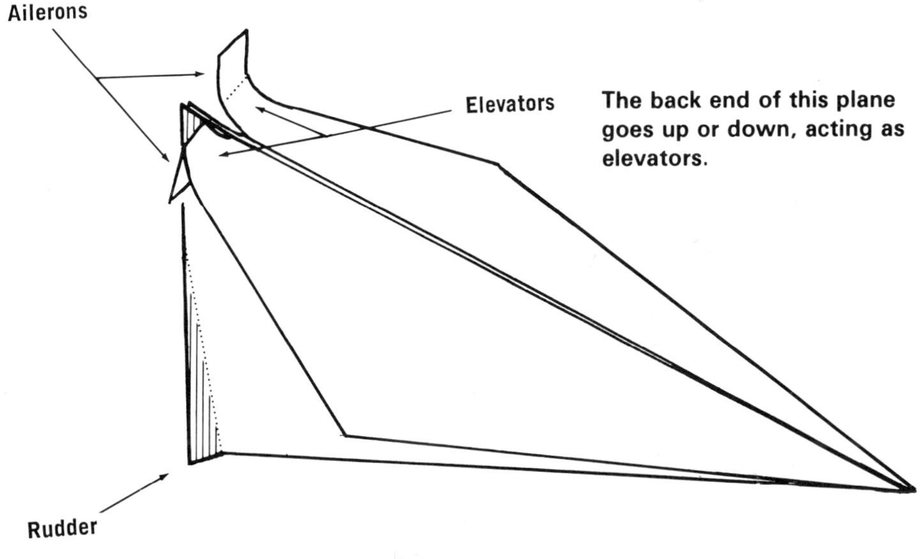

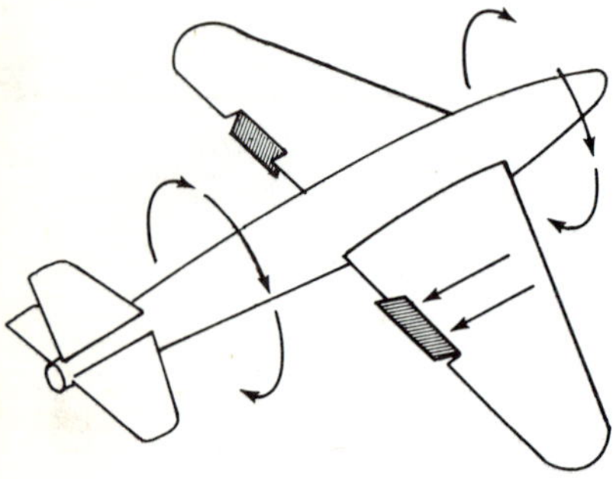

AILERONS control the rolling of a plane. If the right aileron is up, the moving air pushes that side down and the plane rolls over on its side, to the right.

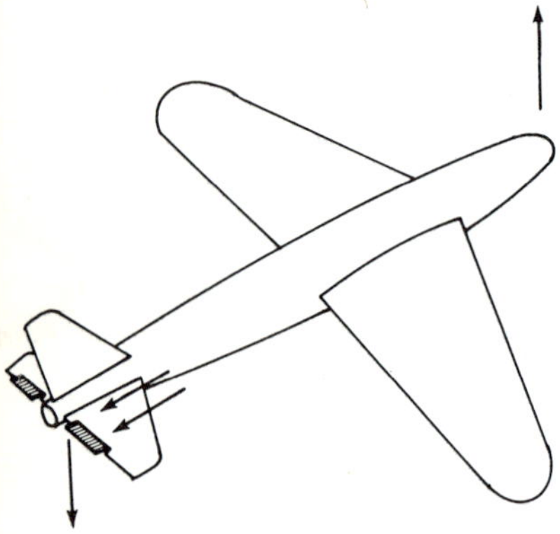

ELEVATORS control the pitching motion of a plane. If the elevators are up, as the plane moves through the air, it pushes against them, and the tail is forced down and the nose goes up.

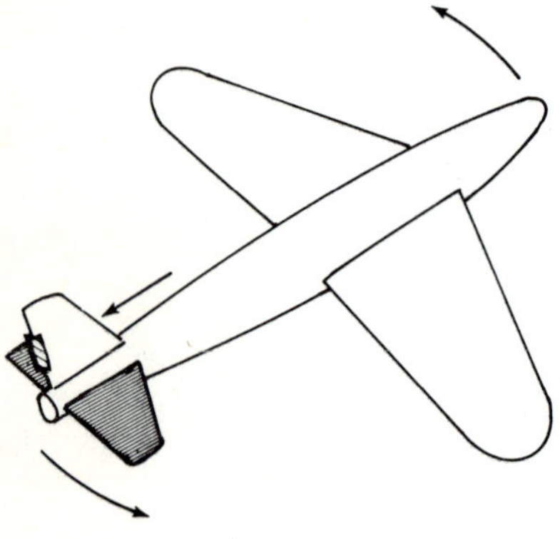

The RUDDER controls the yawing motion of a plane. If the rudder is moved toward the left, air pushes against it, forcing the tail to the right and nose to the left.

Make a model of the AILERONS by cutting a strip of heavy paper to this shape, or one like it and then attaching it to a nail as shown before. Next cut the ailerons and fold them: one side up and one side down. Blow from the front as if these were the wings of the plane. Which side does the paper roll over on? You can make it do the opposite by changing the ailerons. Sometimes a plane is affected by something called Adverse Yaw which means the plane rolls over to the opposite side and this is why planes have rudders *and* ailerons to help them turn.

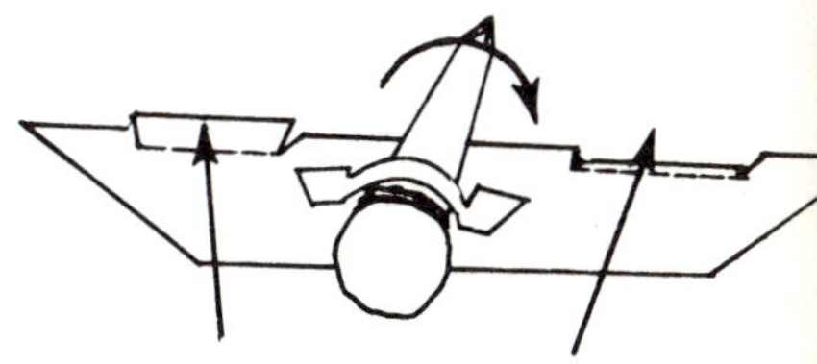

Blow straight at it.

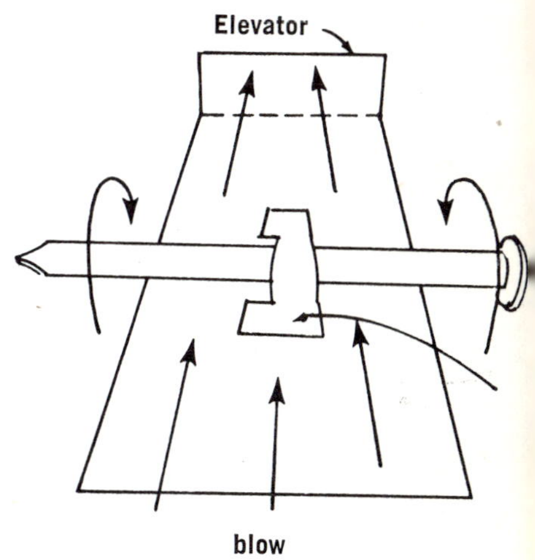

Make a model of the ELEVATOR by cutting a strip of paper or tagboard and folding one end up. Find the balance point and attach a nail or smooth pencil across it so that the paper can move up or down, but not too easily. You can blow from the front and make the nose of the plane go up or down by changing the elevator.

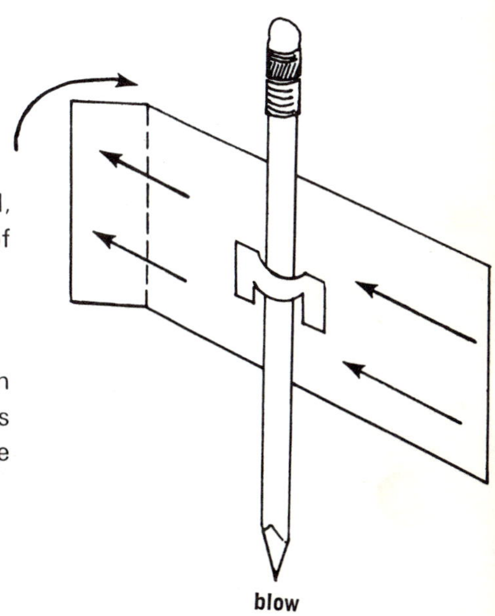

Make a model of a RUDDER. Cut it out of heavy paper, tagboard, or cardboard. You can use the one on this page, or design one of your own.

This rudder is made to be attached sideways to a nail, either by an extra tab of paper cut and taped around the nail, or by cutting slits in the paper strip and slipping the nail through them. You will have to balance it to find the best place to attach the nail.

Photograph by Anthony Galvan III

Photograph by Anthony Galvan III

Some books that will be useful in a study of balsa gliders

Bishop, Richard, *From Kite to Kitty Hawk*, New York, Thomas Crowell, 1958

Cooke,, David C., *Who Really Invented the Airplane*, New York, G.P. Putnam and Sons, 1964

Corbett, Scott, *What Makes A Plane Fly*, Boston, Little Brown and Co., 1967

Feavolo, Rocco, *The Junior Science Book of Flying*, Champaign, Illinois, Garrard Publishing Co., 1960

Kettlekamp, Larry, *Gliders*, New York, William Morrow and Co., 1961

Lopshire, Robert, *A Beginner's Guide to Building and Flying Model Airplanes*, New York, Harper & Row, 1967

Pudney, John, Editor, *Flight and Flying*, New York, David White, Inc., 1968

Snyder, Don E., *Danny's Glider Ride*, Chicago, Follett Publishing Co., 1968

Soaring Society of America, *Joy of Soaring*, 1969

Photograph by Anthony Galvan III

Photograph by Kay Galvan

Part Three

Air and Forces Acting on Planes and Gliders

AIR . . . .

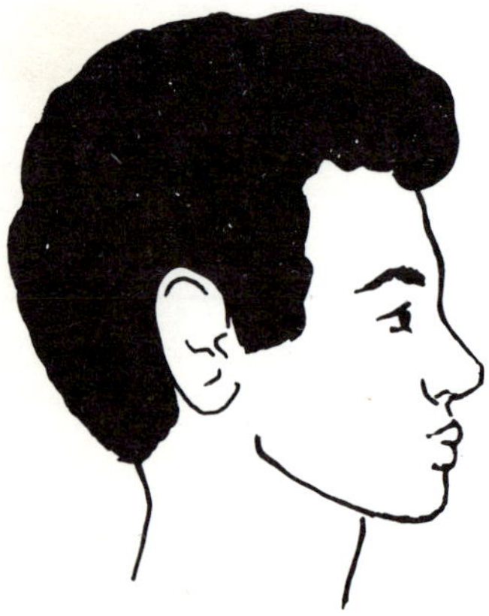

Since science experiments and activities were such a big part of studying gliders, I have made this section of the book about nothing but air and the forces acting on a plane or glider.

What is air? You can't really see it, so how do you know it is there?

One way you know is because you are breathing it right now. You must breathe air to live.

But you can find ways to see air and that will make it easier to know it's there.

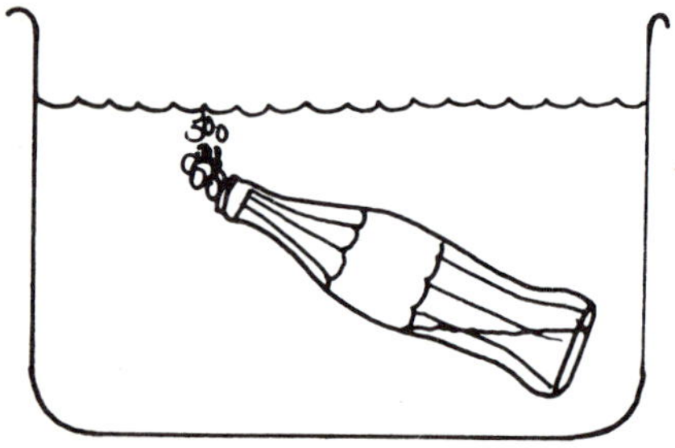

You can see air under water, because it floats up to the top of the water in bubbles . . . air bubbles.

Take a coke bottle full of air; put it under water and pour it out. (Remember, air pours up), then you will see bubbles of air. As the air moves up, the water moves down into the bottle.

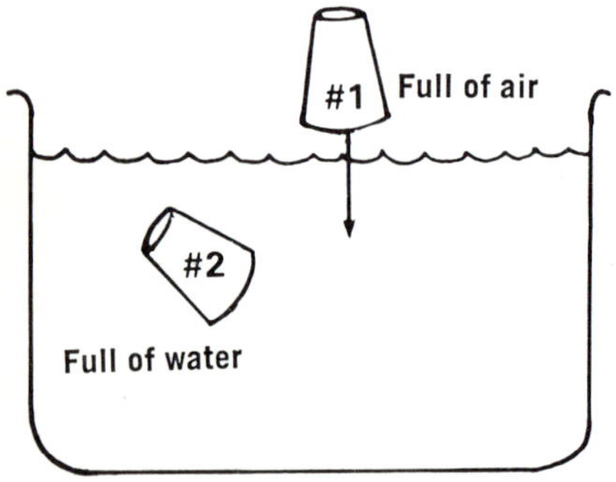

Next time you are washing dishes (or just before you really start) try pouring some air from one glass to another.

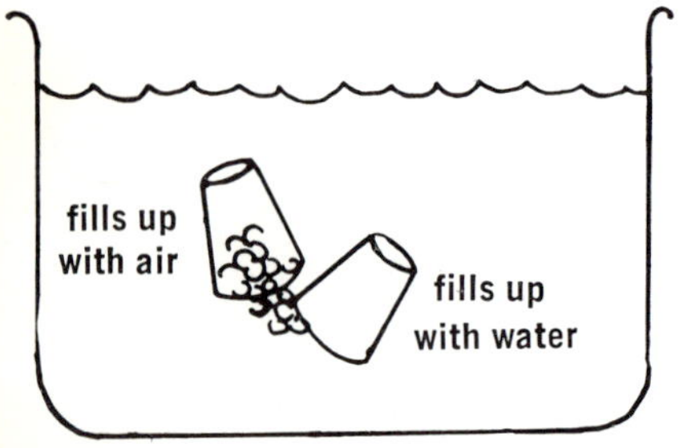

Hold glass #1 upside down full of air. Hold glass #2 in the pan; it will be full of water. Push glass #1 straight into the water and then tilt it and let air pour upwards into glass #2. The bubbles of air will go right into the glass of water and push the water out! Now you are holding glass #2 upside down full of air; try pouring it back again.

Both Dr. David Hawkins and Barry Kluger suggested I include this experiment about the air siphon. It is tricky to do because of needing two people to work in one big tub of water, as well as needing to keep the water out of the siphon hose, which means that you will need large containers that your hands and fingers can fit into, along with the hose.

After you open one end of the siphon to the air, it will travel through the tube and into the other container. Water will siphon down, but air will siphon up.

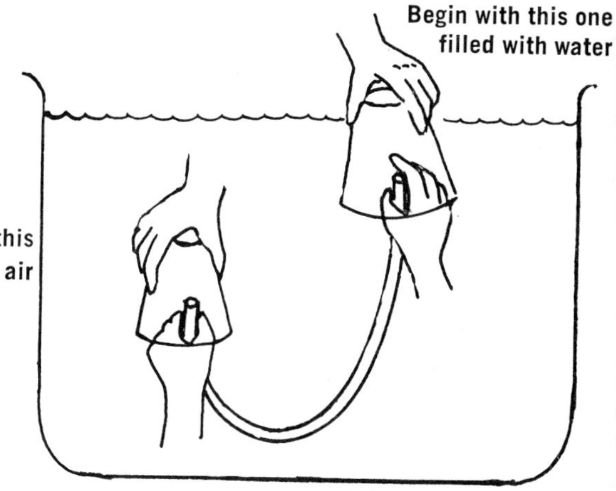

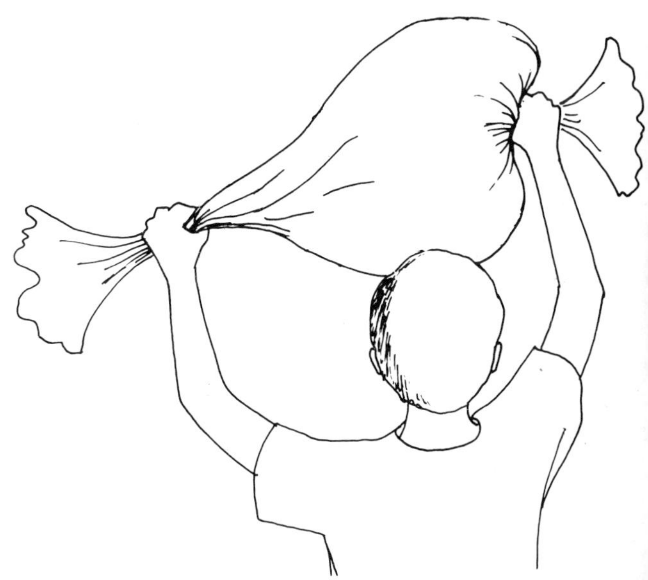

Here's a nice way to play with air. Swish a bubble of air back and forth in a plastic clothes bag from the dry cleaners. Just fill it with a little air and hold the ends tightly, then swish from side to side! Try it up and down. (If you had a bag of water, the water would fall down to the bottom . . . . . where does the air fall or does it?)

Another way you can see air, besides seeing the air bubbles, is by seeing the space it fills up. Watch a parachute full of air.

You can't see air, but you know it is around you because you can breathe it and because you can see air bubbles and because you can see the space it takes up when it is filling something that had no air in it.

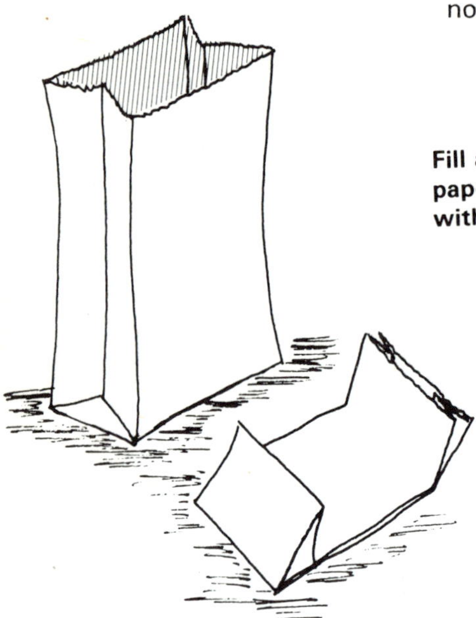

Fill a paper bag with air...

...or a picnic blanket.

Air has weight.

Try this experiment with two ballons on a yardstick.

We tried to show this on a smaller balance and it didn't work very well. Air just doesn't weigh much at all. The demonstration seems to work better with bigger balloons. Can you think of why? Another question that's fun to think about is this: why wouldn't this same demonstration work with a paper sack full of air on one end and a sack without air on the other? Think about a sealed baggie full of water floating in a tub full of water and another empty baggie in the tub.

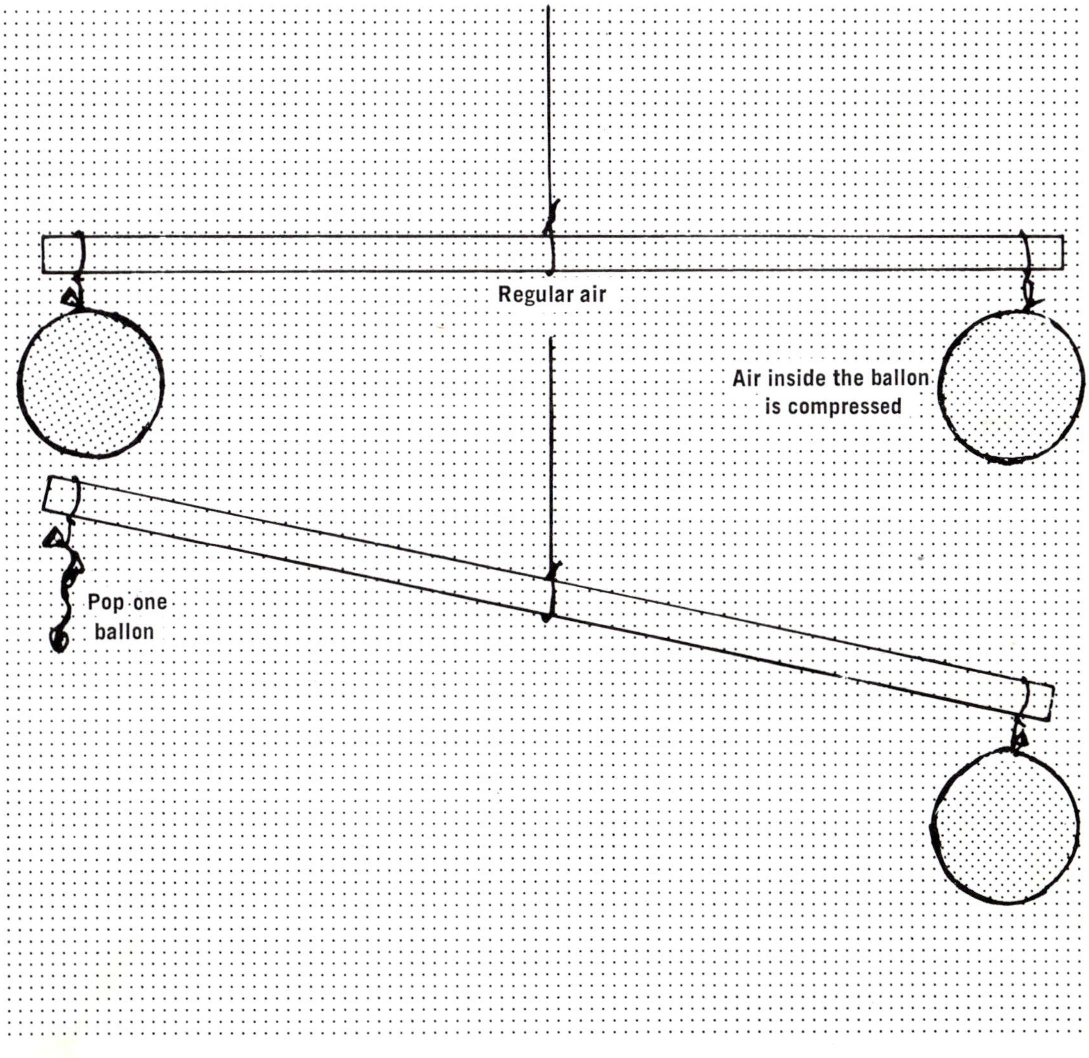

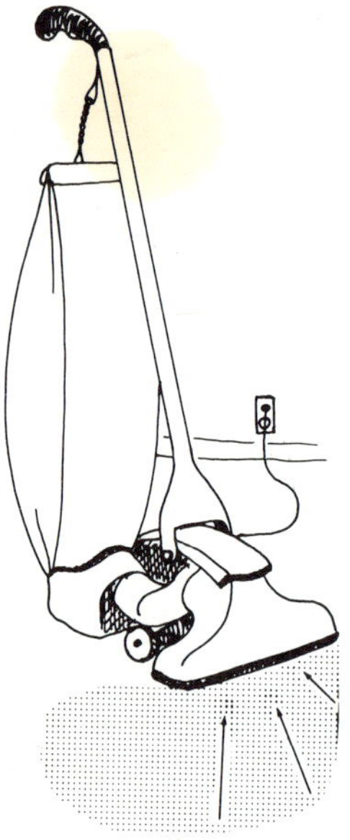

Two other ways to tell that air is all around you are:

1. Air moves. Moving air is wind. Moving air carries things with it, like a kite.

Moving air in a vacuum sweeper will sweep dirt and dust off the floor.

2. Air presses. A fish under water in a lake has water pressing against it from all sides. Air acts the same way on us as the water acts on the fish. This is called air pressure.

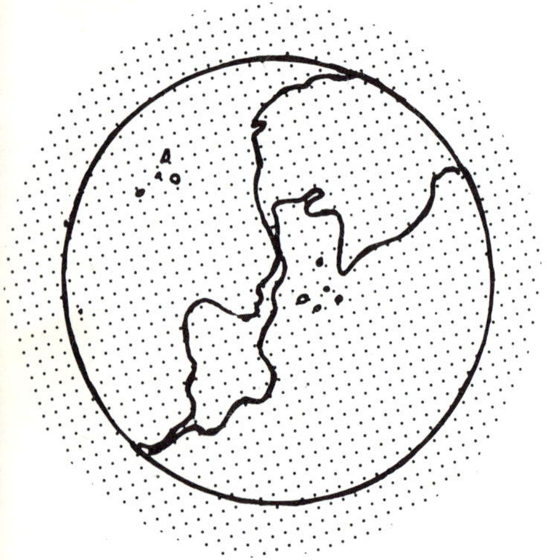

The earth is floating in a sea of air as it says in Wm. P. Gottlieb's book, *Aircraft and How They Work.*

Air is piled all around the earth and under and over it too. That air is resting and pressing on everything. It presses sidewards and upwards as well as downwards.

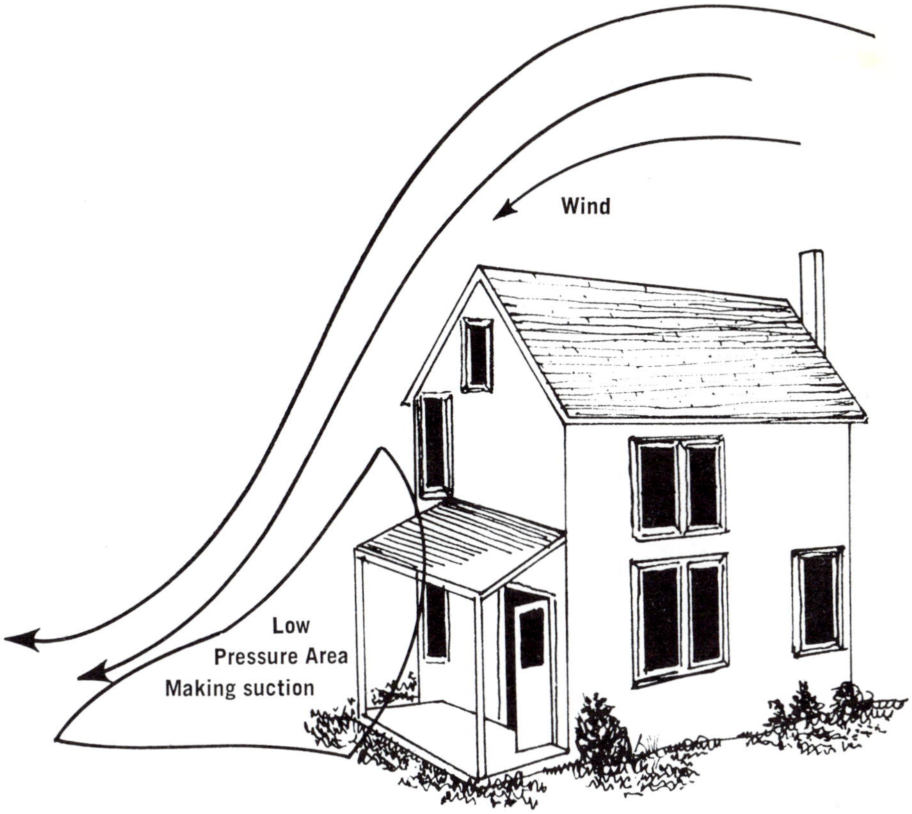

One example of air pressure is something that happens once in a while on a windy day:

The wind might be blowing past your house, creating a low pressure area on the side where the door is.

If someone happens to open the door, another door somewhere in the house will bang shut!

The wind was not inside your house. But when it blew past the house, it made a low pressure area right near the door. As soon as the door was opened, the air in the house was drawn out to fill in the low pressure area and as it was drawn through the house it slammed the other door.

Here is an experiment with air pressure. You can use an empty tin can with a spout, or something like it.

Fill the can with only about one inch of water and heat until it is boiling. As it heats, the air heats, too. Warm air takes up more space than cool air. So some air moves out of the can as it is heated.

Next turn off the heat and cap the hole tightly.

Now wait.

While you are waiting, the steam inside will be cooling and condensing — or turning back into water. But since there is a cap on the hole, no air can get back in to fill up the space that the steam took up.

That means there will be a partial vacuum. The air is not pressing the same on the inside and outside. The outside will be pressing more.

That air on the outside will squeeze and crumple the can until the pressures are nearly equal again. Can you think of some reasons why pouring cold water on the sides of the can will make it crumple sooner?

Try the same thing with a plastic bleach bottle. Heat up the air inside by pouring boiling water into it and then out again. Then cap it tightly. When it cools, the plastic bottle will crumple.

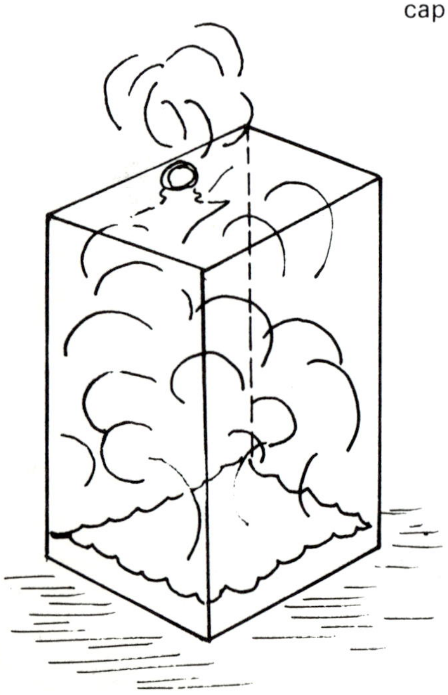

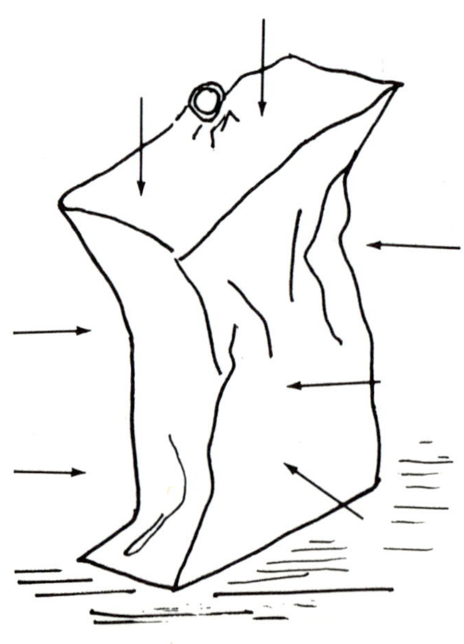

The same type of experiment showing air pressure at work can be done with a coke bottle, pan of water, a hot plate and a balloon.

Fill the coke bottle with hot air. (Pour boiling water into the bottle and then out again. It will be full of hot air.)

Then quickly put a balloon over the opening.

Now watch the balloon. The outside air moves in to fill up the space inside the bottle and it blows up the balloon.

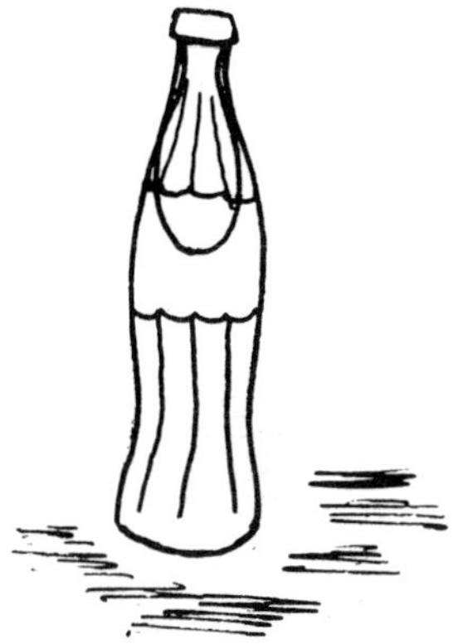

Air pushes and presses from all sides. That is why you can have a cup full of air when it is upside down.

Air also pushes and presses on planes and gliders. As long as the plane or glider is still, the air is pressing equally on all sides. But when the plane moves through the air, the shape of the wings makes the air travel faster over the top of the wing than on the bottom and this causes an area of low pressure over the top of the wings. The plane then moves up into the low pressure area and is lifted up into the air. This is the most commonly accepted theory of lift in aerodynamics. However, a new, possibly more effective way of creating ift has just recently been discovered by Mr. Richard Kline of Atlanta, Georgia. He patented his design for a paper airplane in the spring of 1973. There is a small article telling about this man and his work in the April 2, 1973 issue of *Time* Magazine. More information should be coming out about this in the near future.

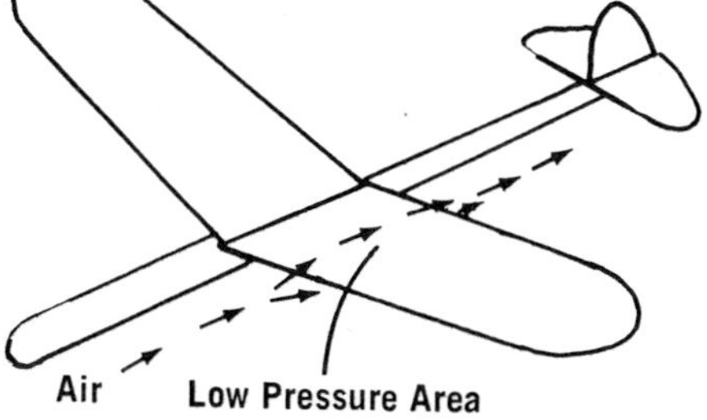

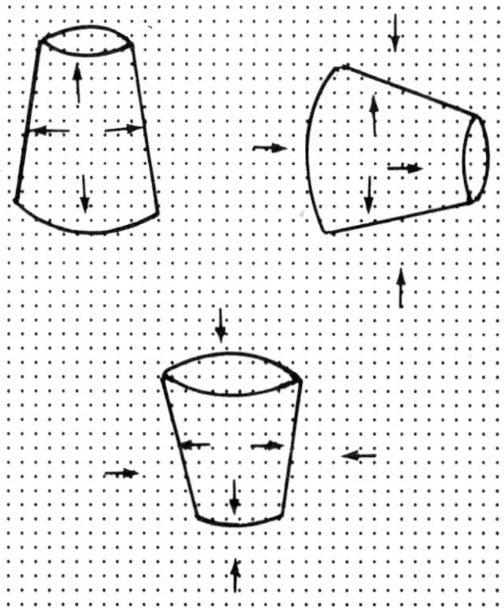

FORCES ACTING ON PLANES AND GLIDERS ....

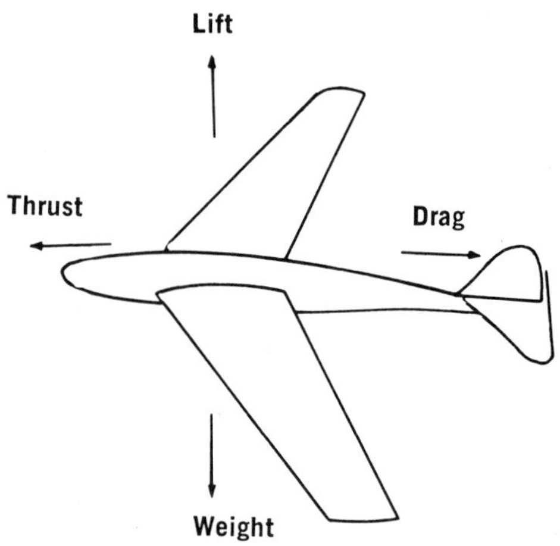

Lift, Weight, Drag and Thrust

As a plane or glider flies, the four forces shown above are affecting it. LIFT is the force pulling the plane up. THRUST is the force pulling the plane forward. DRAG pulls the plane back. WEIGHT (or gravity) pulls the plane down.

In order to fly, the plane or glider has to have more LIFT than WEIGHT and more THRUST than DRAG. The force of LIFT has to be stronger than the pull of the WEIGHT, or the plane would stay on the ground. The force of THRUST has to be stronger than the force of DRAG, or the plane will not move forward.

But if there were no WEIGHT, the plane could be lifted so easily that it would fly through space and keep going on and on! If there were no DRAG, the plane could keep moving forward and not slow down or turn. So all the forces affecting the plane are important. They each can be used in a helpful way.

Lift

To try and understand the force, LIFT, we had a "Bernoulli Day" at school. Mr. Bernoulli lived in France in the 1700's. We did experiments to learn about Bernoulli's principle. which is: the faster air moves, the less pressure it has, or to be simple, the less it can push or press.

Hold a paper up to your chin and blow over it.

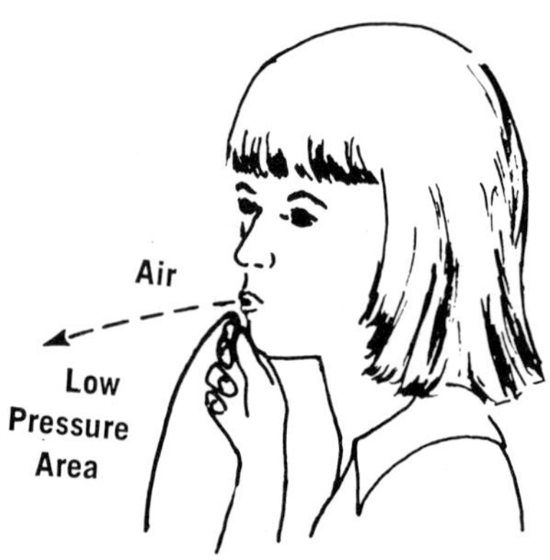

As you blow, the air moves faster. The air can't press or push as much when you are blowing as it was when it was still; it can't press as much as other air that remains still.

Low Pressure Area

That space where the air can't press down as much as everywhere else, is called a low pressure area.

A low pressure space is not really a hole in the air, but it's like a hole in the air because the other air from around it will be moving into that space. The paper under your chin will move up into that space because it goes along with the air moving up in there.

Low pressure areas in air can be a little bit like sugar in a sugar jar: when you pull out a cupful of sugar, all the sugar around it moves in to take the place of the amount you've taken out. You couldn't leave a hole of nothing in a sugar jar and there can't be a low pressure space in the air for very long, because other air will always be moving into that area.

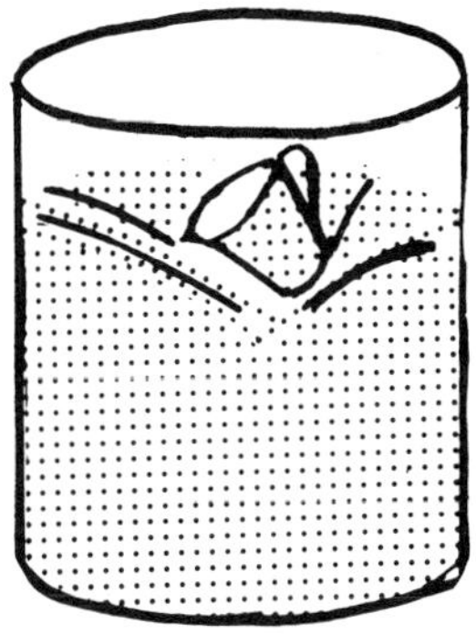

Take two papers and hold them hanging edgewise in front of your face. Now blow right between them.

The faster the air moves between the papers the less it can push. But the air on the other side of the papers and everywhere else, is still pressing just as hard. Because of this, the papers move into the space with the least pressure.

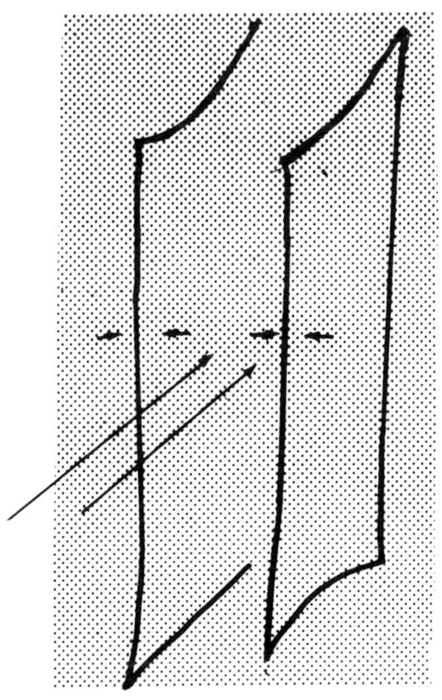

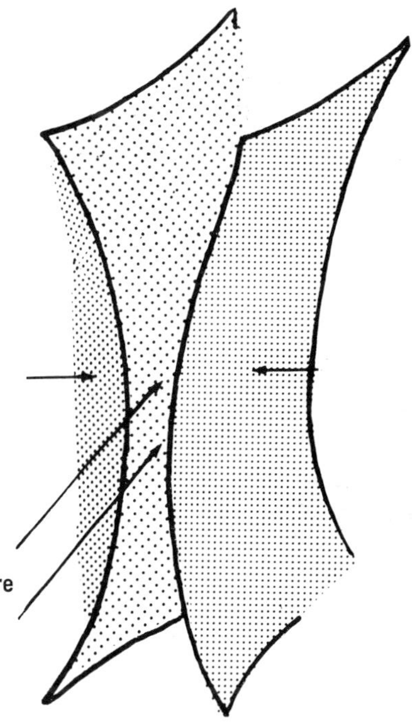

Hang some ping pong balls on strings and tape them so they dangle from a table; then blow between them.

The ping pong balls will move together. The faster you blow the air, the less it can press and the balls move into that space where the air used to be pressing.

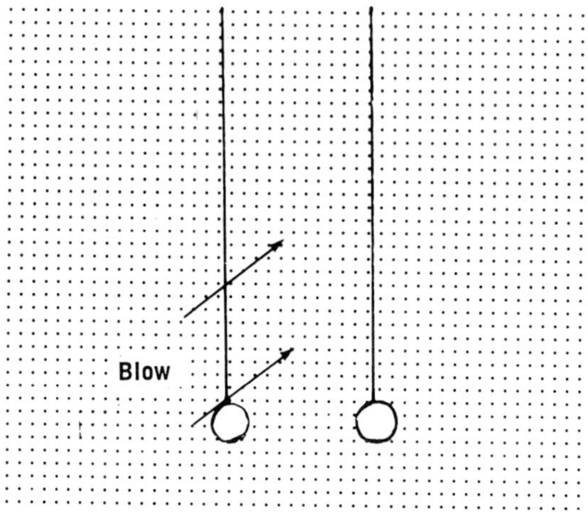

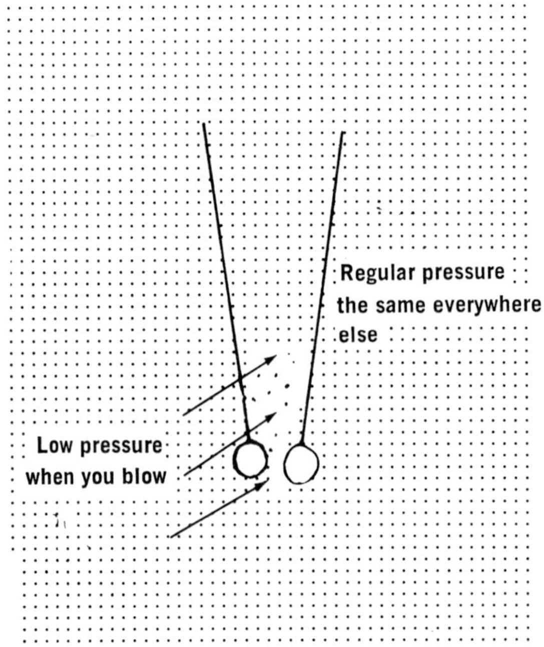

Fill a glass half full of water. Put a tube or clear straw in it. Now blow across the straw or tube through another tube. Water moves up into the straw because of the low pressure in the straw.

The water level rises in the tube in the glass to fill in the low pressure space.

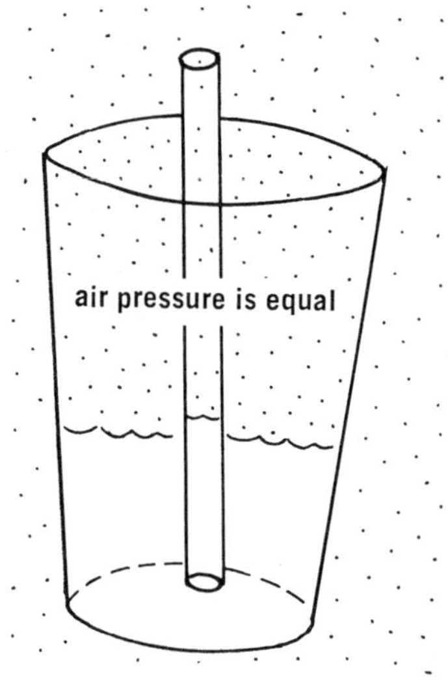

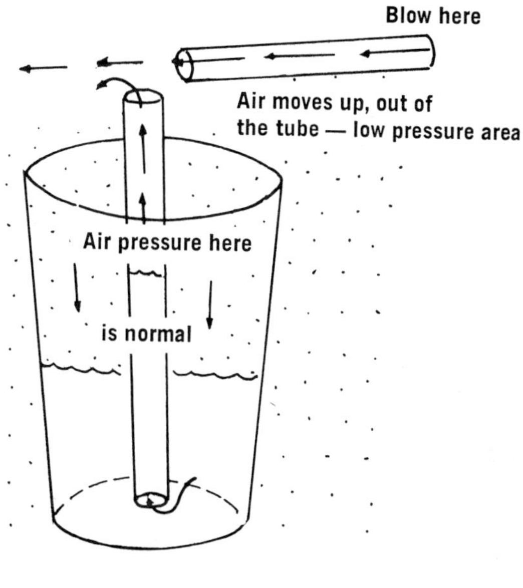

Bradley tried this experiment from *Mr. Wizard's Science Secrets* by Don Herbert (Hawthorne, 1952. p. 64).

Cut a tagboard or light cardboard into a 2 inch square. Stick a pin into the middle of it.

Tip your face back and aim the cardboard at the ceiling. Try to blow it off the spool. Can you figure out where the air is moving the fastest?

Do you think the experiment would work with a larger hole in the spool? A larger hole and a larger cardboard?

Do you think it would work if you tried it upside down? (Bradley asked this question but he was already pretty sure of the answer before his friends tried it.)

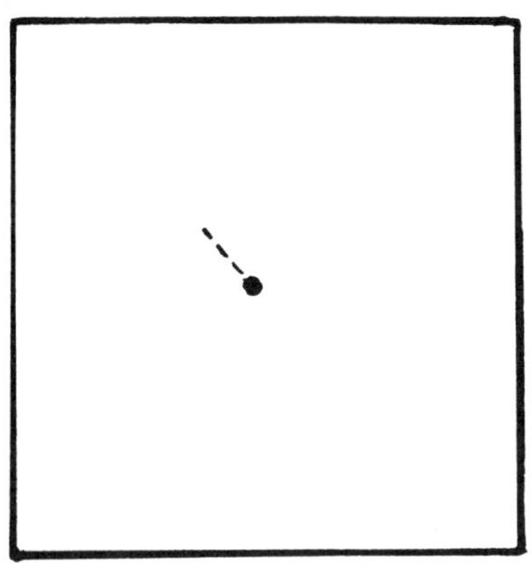

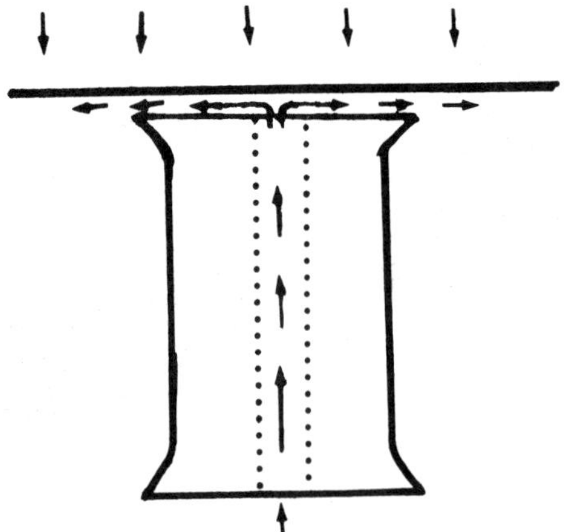

A vacuum sweeper blowing outward will make a low pressure space as long as it continues to keep blowing. If you throw ping pong balls into the space, they won't fall out or fall down because there is air under normal pressure all around the sweeper which is flowing into the space.

This vaccuum sweeper demonstration is difficult to do well, because there is a lot of turbulence created when the air rushes through the sweeper; so the longer the hose, the better. That way the air stream can flow more evenly and will not create a lot of bubbles and pits which end up bouncing the balls around and out.

The other problem is that you have to get a sweeper that you can reverse the air flow on, so it is blowing instead of drawing.

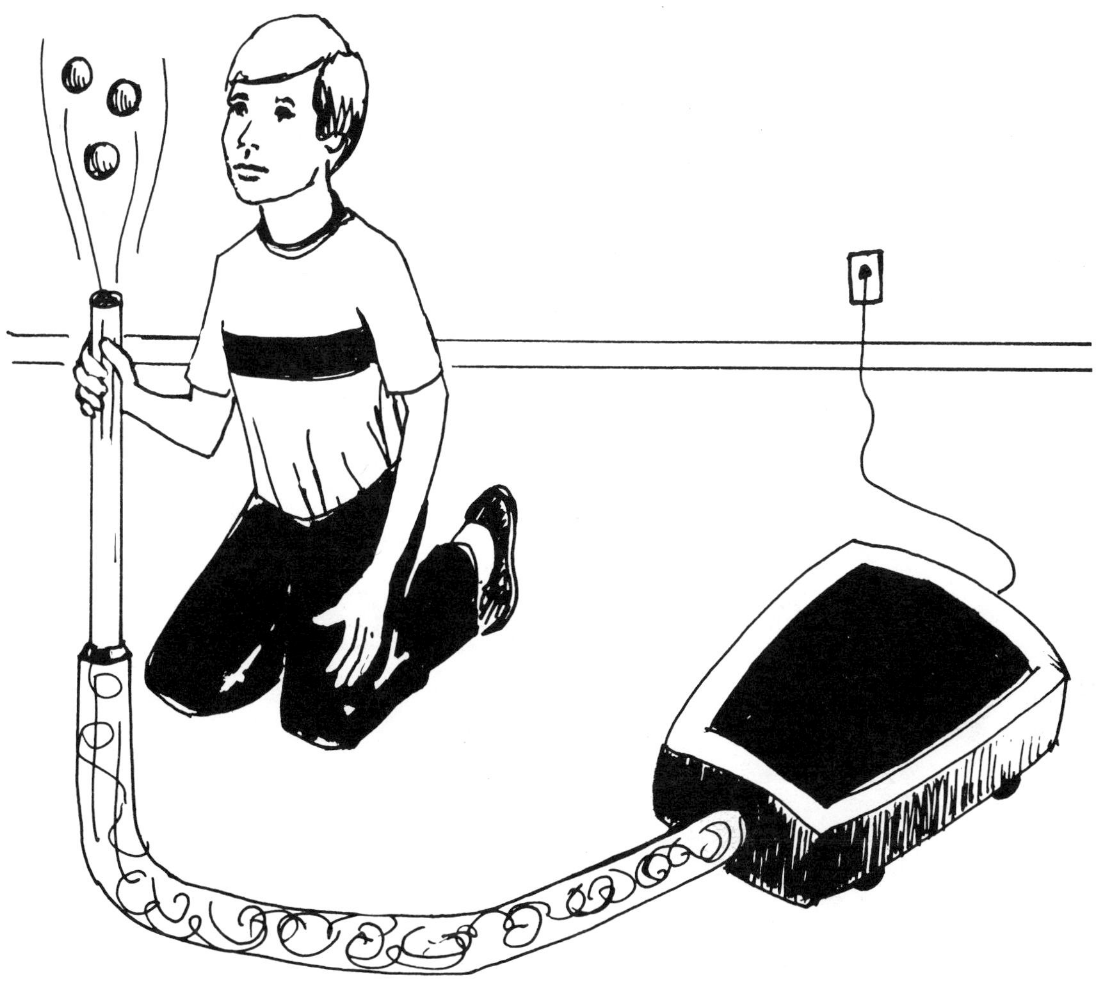

Airfoil

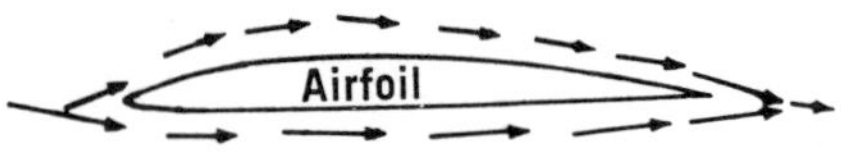

A cross section of the shape of an airplane or glider wing is called an airfoil. This airfoil is another working example of Bernoulli's Principle.

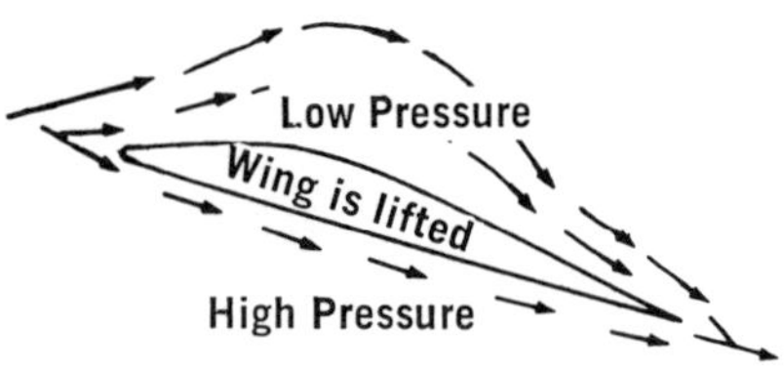

Since the top side of the wing is longer than the bottom (measure with a string if you aren't sure) the air has to travel faster across the top to meet the bottom air at the back of the wing at the same time.

The faster the air moves, the less it can press down on the wing, so the wing moves up into the low pressure area. This is the main way a plane or glider gets lifted up into the air. It also gets some lift from being pushed up by the air underneath, but not as much.

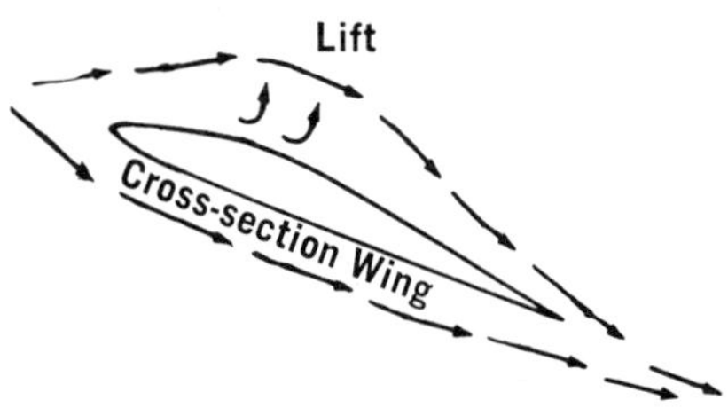

As the wing lifts up, it meets the air at a certain angle. This is the ANGLE OF ATTACK.

Bernoulli't Principle only works when the wing is at a good angle of attack.

If the wing is at too steep an angle of attack, the air on the top of the wing will not move fast enough to meet the bottom air, and LIFT the wings.

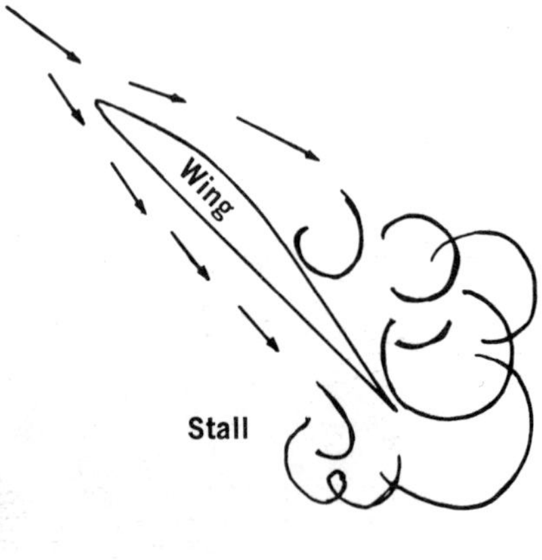

Instead it will tumble and roll like water over a waterfall and all the TURBULENCE this makes, will drag the plane or glider down and make it STALL.

Weight

The force of LIFT must overcome the force of WEIGHT if a plane or glider is to fly. WEIGHT means how hard gravity is pulling on something. If it weighs a lot, gravity is pulling it a lot.

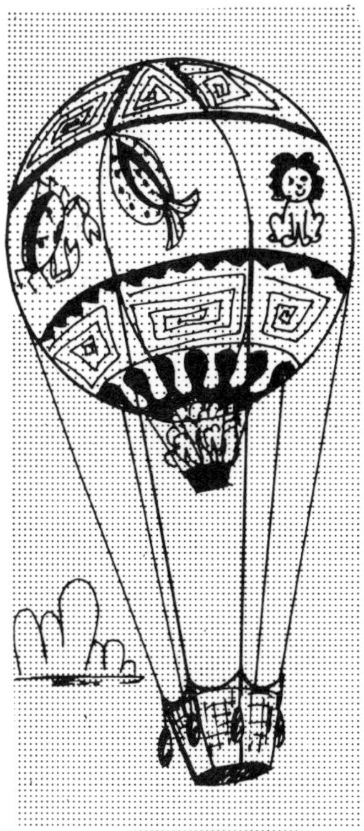

Weight is something that everything has, even air. The amount of weight depends on how close together the molecules are packed in something.

The closer together the molecules are packed, the more the weight, or the heavier something is.

One of the first kinds of balloons to be sent up into the air was a balloon filled with warm air. It was called a Montgolfier, named after the men who invented it.

The molecules in warm air are farther apart than the molecules in cool air. So the warm air weighs less and will float up into the sky.

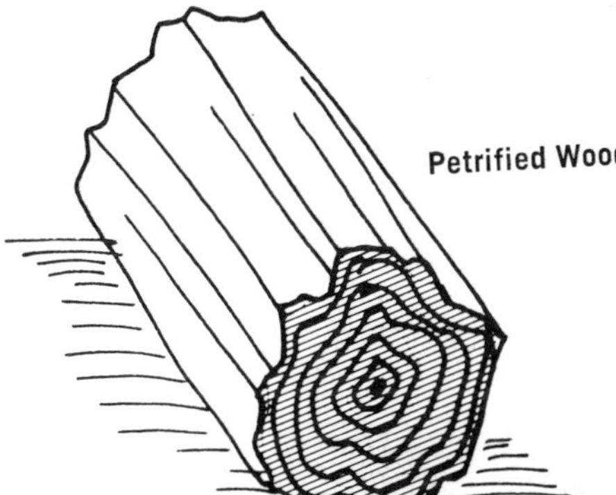

Petrified Wood

There are some things which have molecules very close together. These things weigh a lot. They will sink in water.*

Petrified wood, metals and rocks are some of those things. Some wood, like teak, has molecules so closely packed that it will sink in water, petrified or not!

There are other things which do not have the molecules packed very closely together. Some rocks, like pumice, have so much space between the molecules that air can get in and the rock can float on water.

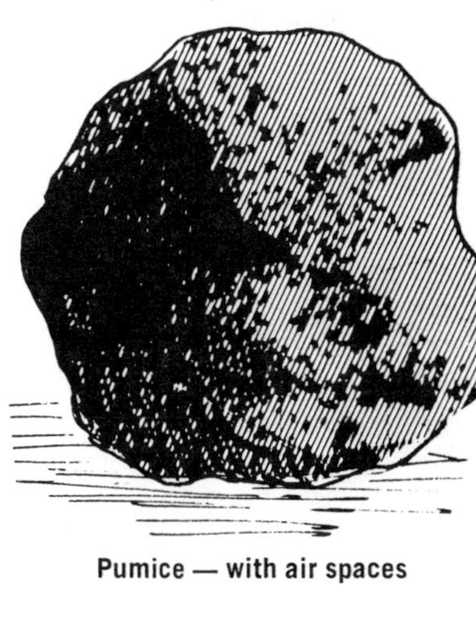

Pumice — with air spaces

*This is simplified. It is possible that there could be two things with molecules packed together in about the same way, but one might weigh more because the atoms which it is made of can be heavier than those of the other item.

Some things sink — other things float — in water. It's the same with air. But most things sink in air. Only a few things float in air.

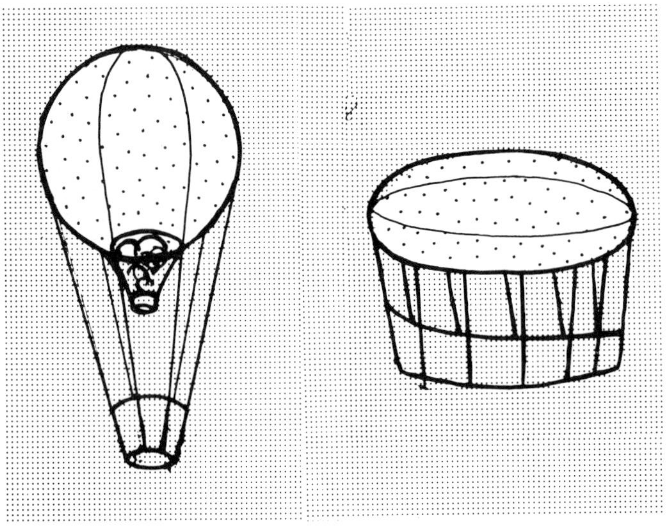

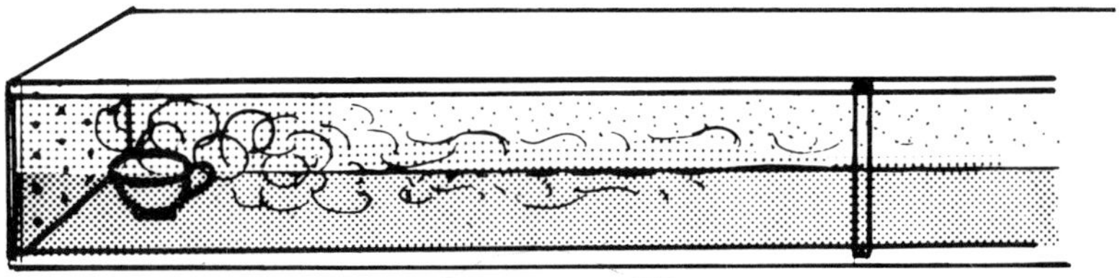

With our wind tunnel, we saw that the cool air of the dry ice in water was sinking and flowing along the bottom of the wind tunnel.

Gravity was pulling harder on the air partly because it was very cool.

But gravity was also pulling harder on the smoke from the dry ice because it was CO^2 (Carbon dioxide), which is heavier than regular air.

A glider or plane has weight too. This weight has to be overcome by the amount of LIFT it has.

One thing that helps lift a plane is speed. The faster the plane goes, the faster the air moves over the wing and the more lift it creates.

Another thing that gives a glider lift is a thermal.

Thermals

A thermal is a bubble of warm air that is rising.

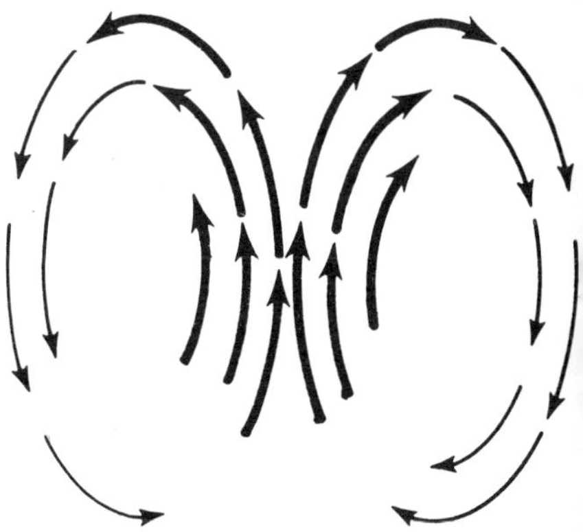

Thermals work the same way the hot air in the Montgolfier worked: the air is warm and the molecules are farther apart so the gravity is not pulling as hard on it. The warm air of the thermal floats in a big bubble up into the sky.

The earth is heated by the sun. Some places are heated more than others. Blacktops and asphalts (such as parking lots) heat up more quickly. Fields heat up more quickly than a park with trees and a lake.

Air is swirling and rising where it is warmer and drifting and falling in places that are cooler. When the warm air rises, the cool air around it will rush in to fill up the low pressure area that it leaves.

Convection Currents

Warm air rising out of a heated space couldn't just leave nothing in the place where it had been. The cool air will always move in to fill up the low pressure space that is left. This movement of the warm and cool air is called a CONVECTION CURRENT.

You can study convection currents and the weight of warm and cool air by making a convection current box.

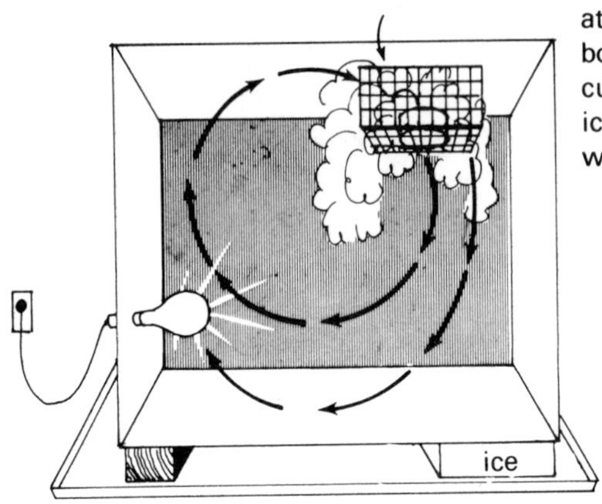

The cold ice under the sides pulls the air down and the light bulb heating up the middle will cause the air to go up. Then, as the warm air rises in the middle, the cool air from the sides will move in to fill up the space. It was difficult for us to get a good convection bov to work, because it was hard to see the smoke. The smoke pellet worked fairly well and we were able to get the idea of the air flow. fill up the space. It was difficult for us to get a good convection box There is also some problem with turbulence from the corners. The experiment would probably work best in a large round drum turned sideways. (Perhaps a 30-gallon drum).

Materials needed for Convection Box

A cardboard box

A 100 watt bulb or a hot plate

Clear plastic wrap to go over the front
Heavy tape

Dry ice or maybe a thick 1" cake of ice frozen in a bread pan,

Smoke pellets,or dry ice in a cup of hot water to make the air smoky to see more easily.

Center of Gravity

The WEIGHT of anything has a balance point. It is called the Center of Weight, or the Center of Gravity. (C.G.) The C. G. is fun to find on different things.

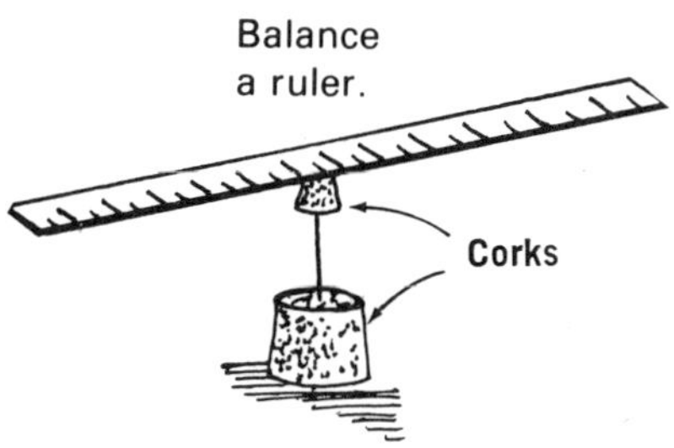

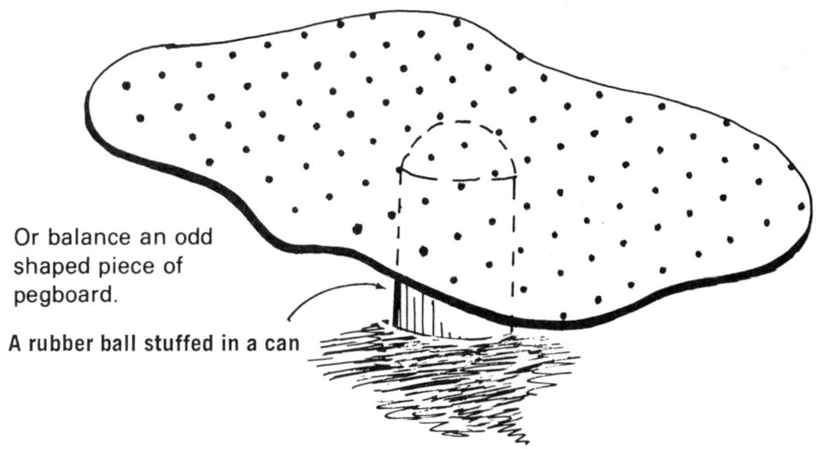

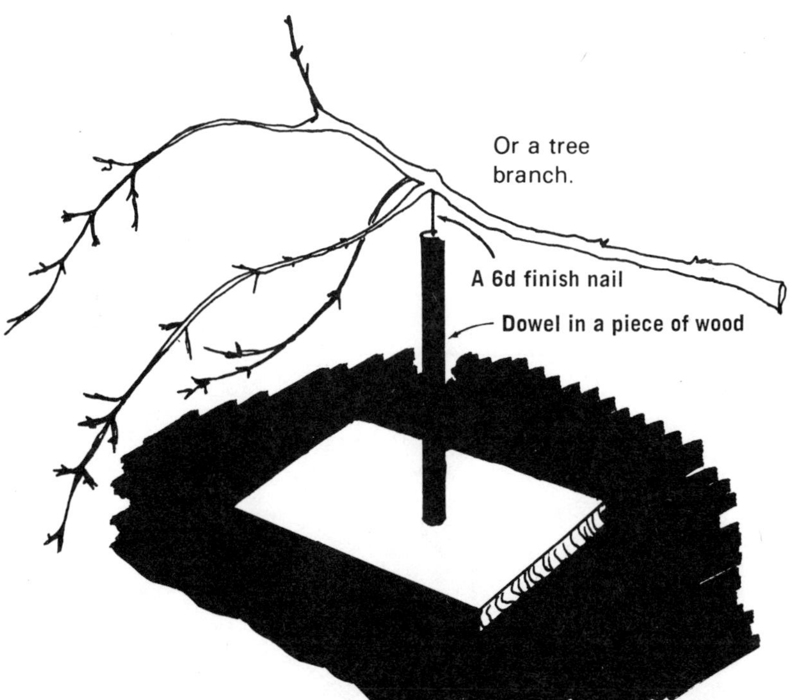

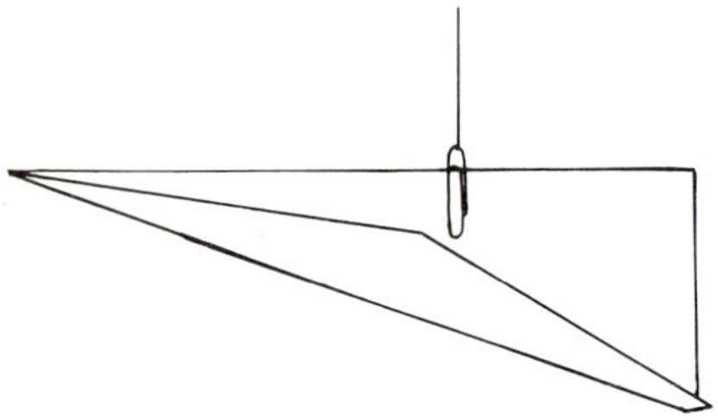

A Center of Gravity can be found on planes, too. It's easy to find it on a balsa wood glider, or a paper airplane.

Find the C.G. on your planes!

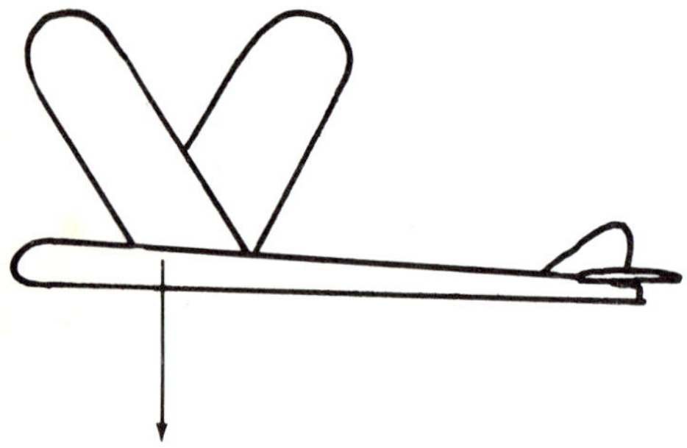

WEIGHT in a plane can be very great. The molecules of steel are packed together very closely. In order for the plane to fly, the LIFT must overcome the WEIGHT.

In order for a plane to fly well, the Center of Gravity and the Center of Lift must be in the right places in relation to each other.

Sometimes in order for your model to fly well, you have to adjust one or the other, or both of them. The C.G. can be moved forward by adding weight to the nose. You can move the C.L. back by adding a control surface on the horizontal tail which is bent down. By trying one thing or another, or a combination of the two, you can usually get a pretty stable glide.

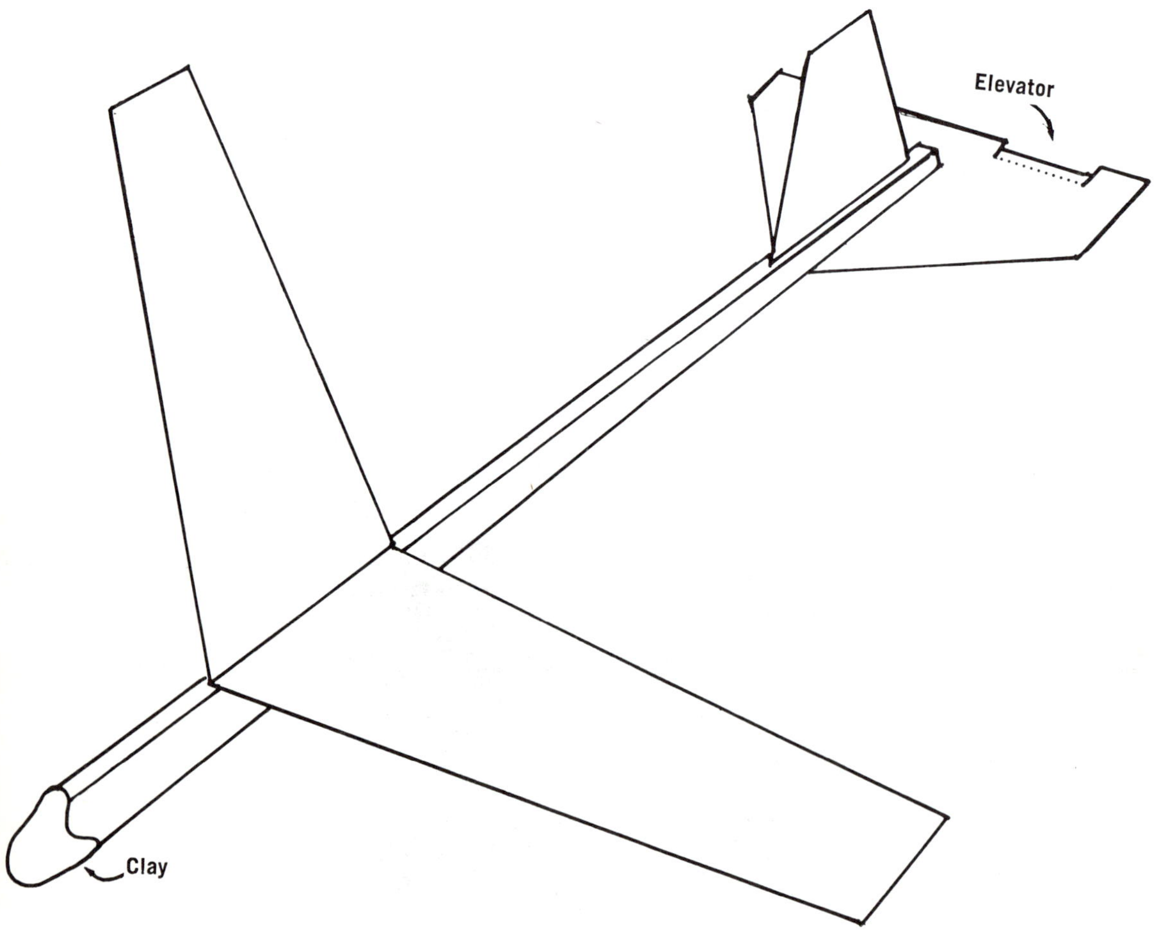

Thrust

In the 1700's a man named Isaac Newton became very interested in the same thing we are studying here. After he watched how things moved, he wrote down what he had seen and kept watching and observing. He wrote three laws of Motion. This is the Third Law: (to put it simply:) For every push one way, there is a push the other way, of the same strength.

Next time you ride your bike and you're going fairly fast, stick your hand out and feel the wind against it. (You will feel the wind in your face too.) Your hand is going the same speed as the bike (which might be 20 miles and hour). As your hand pushes throught the air, it causes the air to push back, at the same speed.

Another way to say it is: the action of your hand going one way, causes a reaction of the air to push the other way.

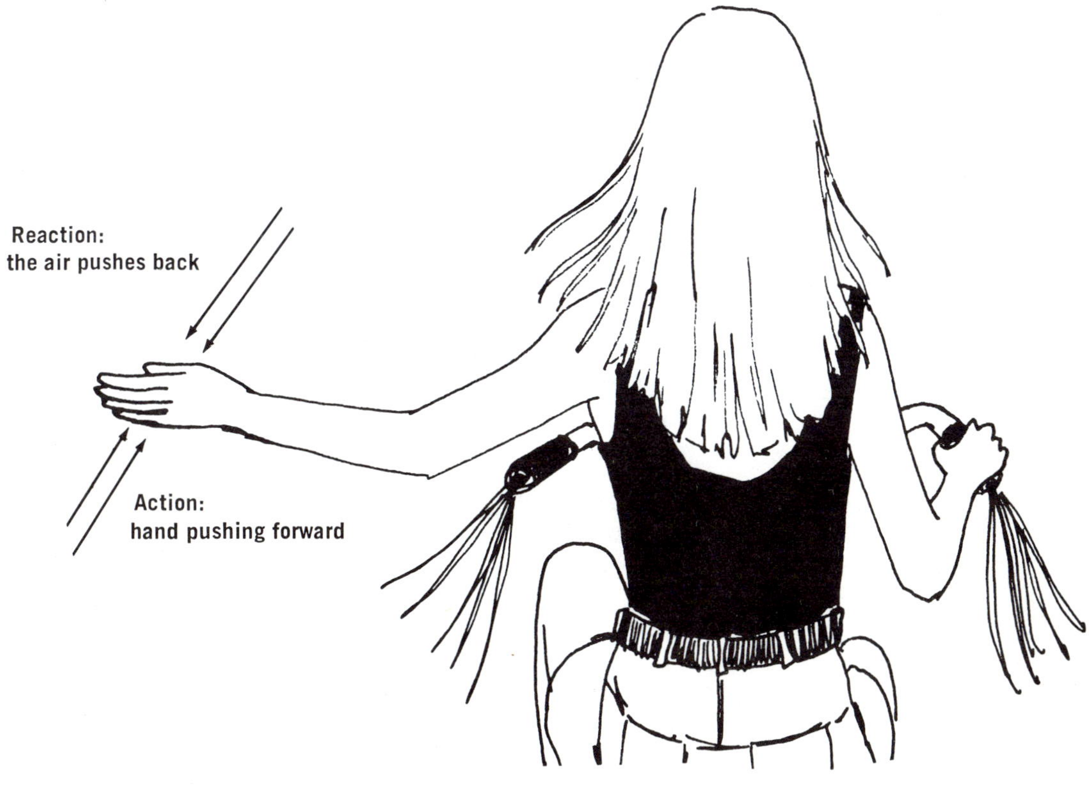

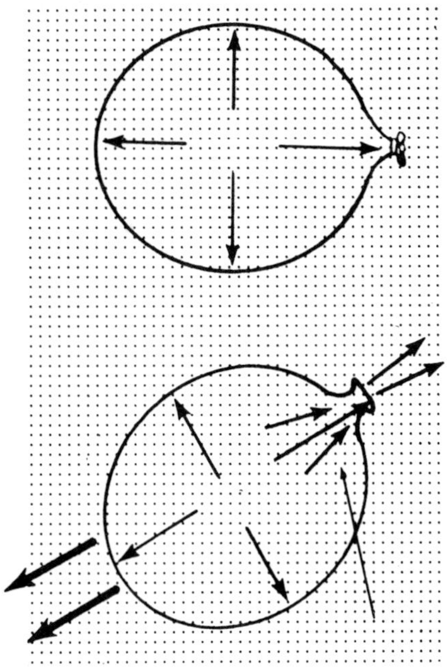

Here is a popular experiment showing action and reaction: the balloon when blown up and tied, won't move. Now let the air out of the opening.

Air inside the balloon is still pushing on the back, but not on the front. So the balloon is pushed back by that air still pressing there, while the air near the hole in front is escaping. The action of the air escaping, makes the reaction of the balloon moving back.

(This is of course an ideal situation with a perfect balloon, with rubber stretching perfectly equal and a spout that stays straight and doesn't wobble. If you have a normal balloon, with a squiggling nozzle and uneven, cheap rubber squeezing unevenly, the balloon will at least make part of a turn in its backward flight!)

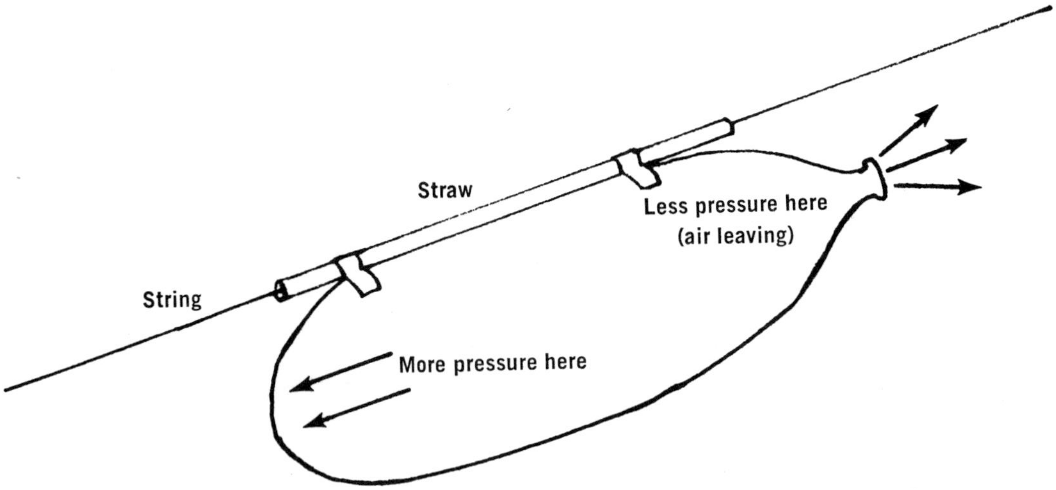

You can experiment with different things to learn how Newton's law works.

Put a fan on a little gym-dollie. The fan blows wind one way, and thrusts itself back the other way (the reaction.) Since it is on the dollie, it will move backward. But if it were sitting still, you would not be able to see the energy of it pushing itself the other way since the earth, and floor, would be pushing back with equal force.

A balloon blown up and let go, will woosh air out one way (action) and move itself backward (the reaction) the other way. The action is always the same strength as the reaction: a reaction can't be stronger or weaker than an action. Every action makes an equal and opposite reaction.

When you jump rope, the same thing happens. There is a push downward and a push upward (from the earth). But we just don't think about that upward push very often, unless one time it isn't there. Have you ever jumped into a snowbank expecting to hit the earth right away, but kept right on going, and *then* hit the earth? It hits back as hard as you hit!

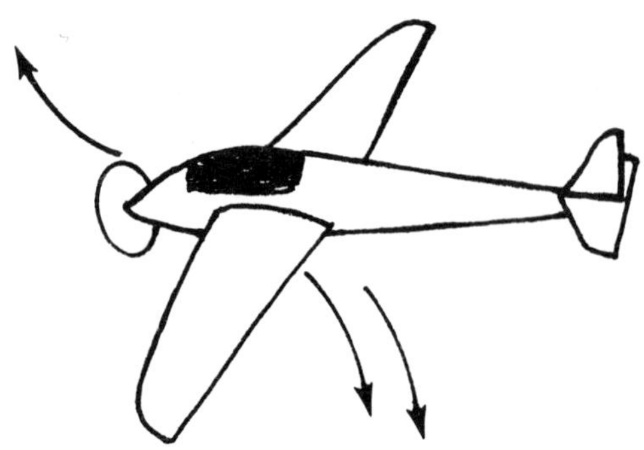

Planes, birds and flying insects have one phenomena in common: they move one way and the air moves back in the opposite direction. Action: the plane or bird moves one way. Reaction: the air moves the opposite way.

We tried this experiment from *Wings In Your Future.* It shows Newton's Law working in still one other way.

You need an Alka-Seltzer tablet (or any fizzie stomach anti-acid)
A small bottle (medicine bottle is about right)
A number of round pencils
A cork to fit the bottle
A long thin slip of paper
Water

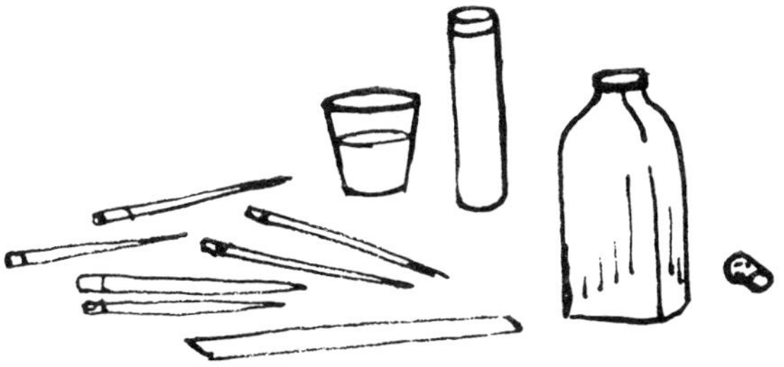

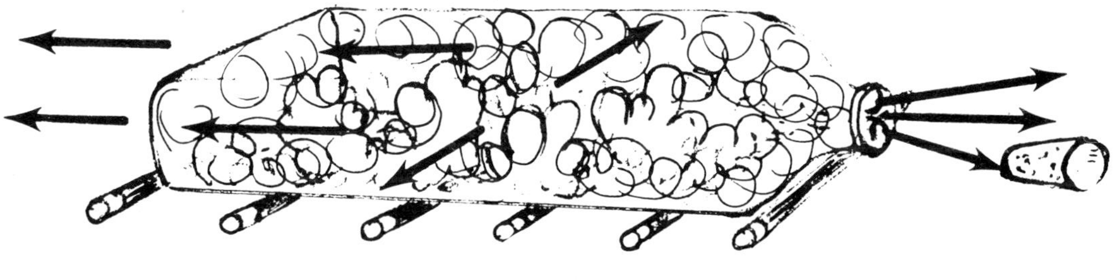

Put about a half inch of water in the bottle and turn it on its side. Then lay it on top of the pencils. Fold the long strip of paper lengthwise and slip it in carefully, with the powder on it and cork the bottle gently. As soon as the water soaks through to the powder, it will make a gas (CO^2) and the gas will press against the insides of the bottle and the cork and make it pop off, in one direction. The bottle will roll back on the pencils in the other direction. We had to be sure the cork was packed lightly and have a sponge handy for mop up!

You can also try this experiment with vinegar and baking soda.
Action: the cork and air escape one way
Reaction: the bottle is thrust the other direction.

Relative Wind

When your hand moved through the air at 20 miles an hour, that was the action. The air pushed back against it. That was the reaction. Even on a very still day, when there in no wind at all, your hand pushing one way at 20 m.p.h. will cause air to push back the other way at 20 m.p.h. This air pushing back is called the RELATIVE WIND.

The RELATIVE WIND always blows against the object that moves against it. Even though there might be another wind that day, going another direction, the RELATIVE WIND will go directly opposite the moving object.

The speed of a plane or glider moving forward causes a RELATIVE WIND (the reaction) to push against itself. The RELATIVE WIND rushing over the wing shaped to an airfoil gives the plane LIFT. If the AILERON is down on one wing, that wing will be pushed up by the RELATIVE WIND.

Thrust

Thrust in an airplane comes from the propellers pulling the plane through the air.

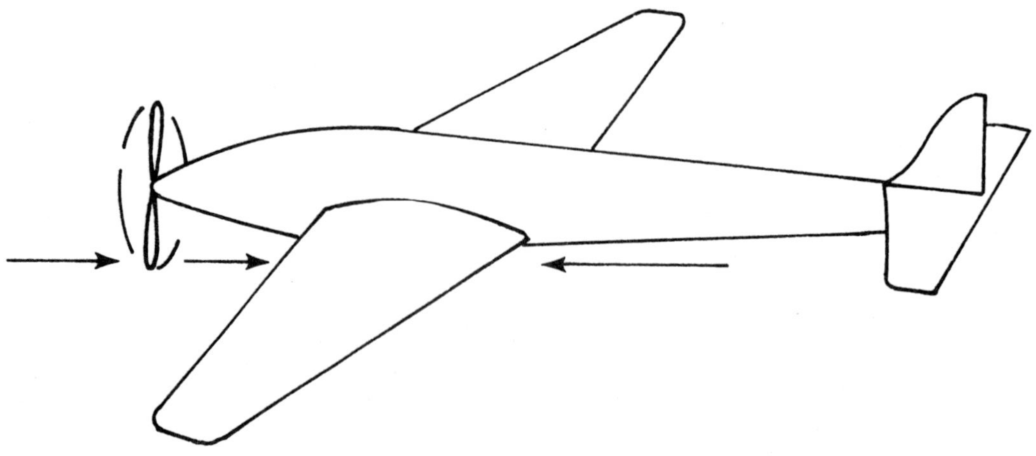

A jet gets its thrust from a different type of power: the jet engine has a powerful explosion of fuel rushing out one way and this causes the jet to be thrust the other way.

Thrust for an airplane or jet is always there. There is power to make it go forward and keep up the RELATIVE WIND and the LIFT. But a glider has only one moment of thrust: the launch. After that, it glides down. It will only go up if it hits a thermal. (And even then the glider is still gliding down, but it is gliding down in air that is rising up, faster than it is gliding down!)

That is why it is so important to shape a good airfoil on your hand launched Glider. The airfoil will give it more chance to lift and fly farther, longer.

Dr. Christopher showed us two good ways to launch our gliders. One was the straight over-hand throw (like a baseball pitch.) It was important to practice the launch a few times because sometimes we would forget to let go of the plane when our throw was at the highest point. Instead, sometimes we accidentally let go when it was at the end of a pitch and that sent the glider straight into the ground.

The other way to launch was to throw sideways. We would hold the model to the side and tilt it at an angle. This launch was also very successful. The plane should be held so the fuselage makes about a 45° angle between the head and arm.

Thrust is pretty simple for a model glider. All the energy comes from your hand and arm.

Thrust for a real glider comes from several different places. The most common way to get thrust (or a launch) is from a tow plane. A powered plane takes off and tows the glider behind it. When the glider has enough speed, it can release the tow hitch and the RELATIVE WIND will lift it and carry it. Then it can slowly glide down.

Cars and mechanical launchers are also used for gliders.

An experienced glider pilot will not only glide on the wind and go down, but will also soar. Soaring is when any plane climbs in altitude without any power. Gliders do it by steering into thermals and circling. The rising air carries them upward and then they glide down to the next thermal and rise up again with it. It is possible to climb many hundreds of feet this way and gliders can travel long distances across the country this way.

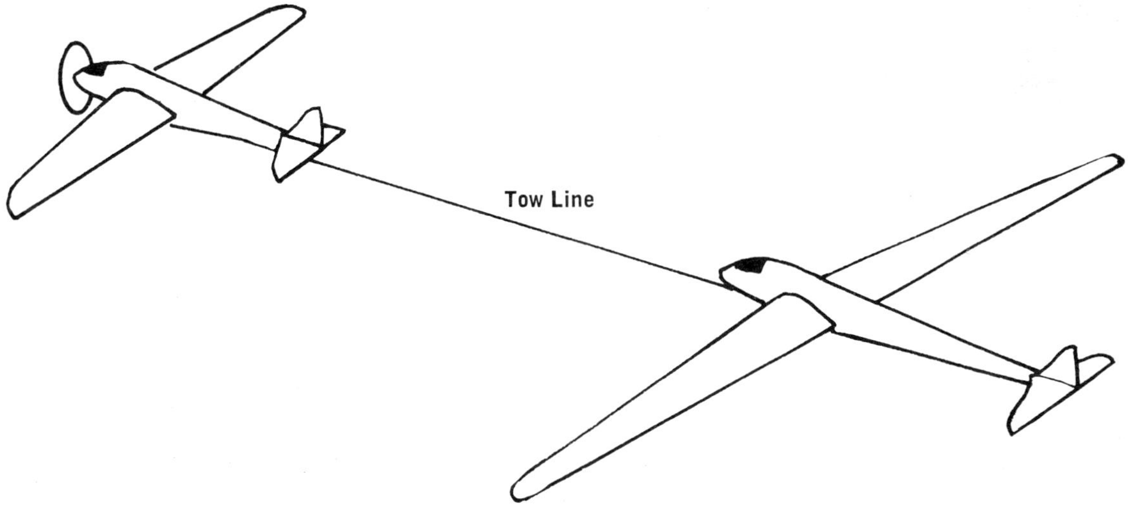

Glide Ratio and Drag

Soaring from one side of the country to the other is done by using thermals and warm air currents to gain altitude. Gliders are built with a high GLIDE RATIO so they can travel far without going down much.

The distance a glider will go down when it glides depends on the GLIDE RATIO. The glide ratio compares the distance a glider moves forward with the distance it goes down. It is comparing the amount of LIFT to the amount of DRAG. DRAG is the force that pulls a plane down and back.

A glide ratio of 12 to 1 means it will go 12 feet forward for every foot down that it goes.

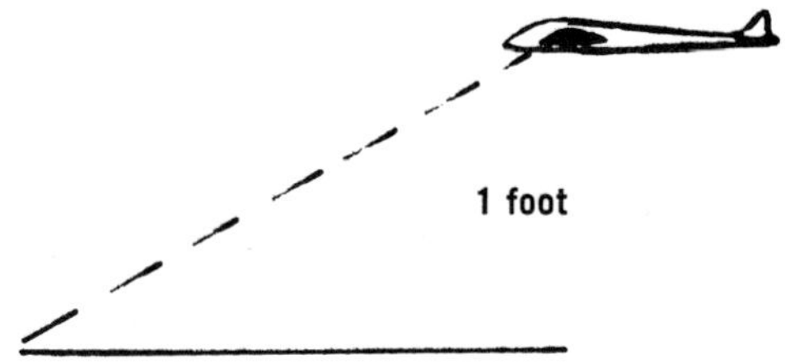

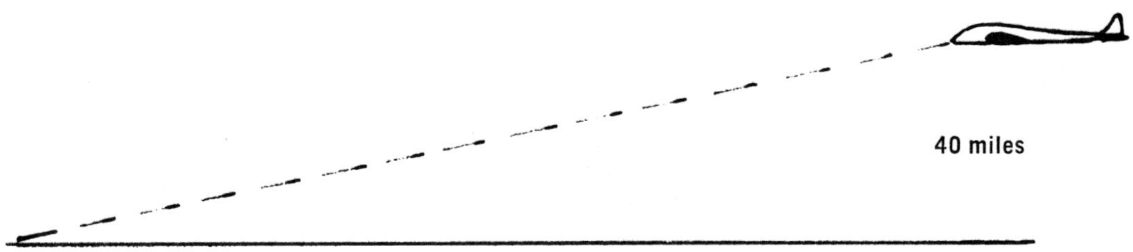

A glide ratio or 40 to 1 means it will go 40 feet ahead for 1 foot down. Or 40 miles ahead for 1 mile down.

A normal glider has a glide ratio of between 25 to 1 and 35 to 1. The extra well built gliders for cross country soaring are built with better ratios of around 40 to 1 or 50 to 1.

Drag

DRAG on a glider for a plane is air resistance. It is the force which acts against THRUST.

The more streamlined a shape is, the less drag there will be. The less streamlined the shape is, the more drag there will be.

A shape that is not very streamlined will create a lot of turbulence behind it and this turbulence will slow the plane down.

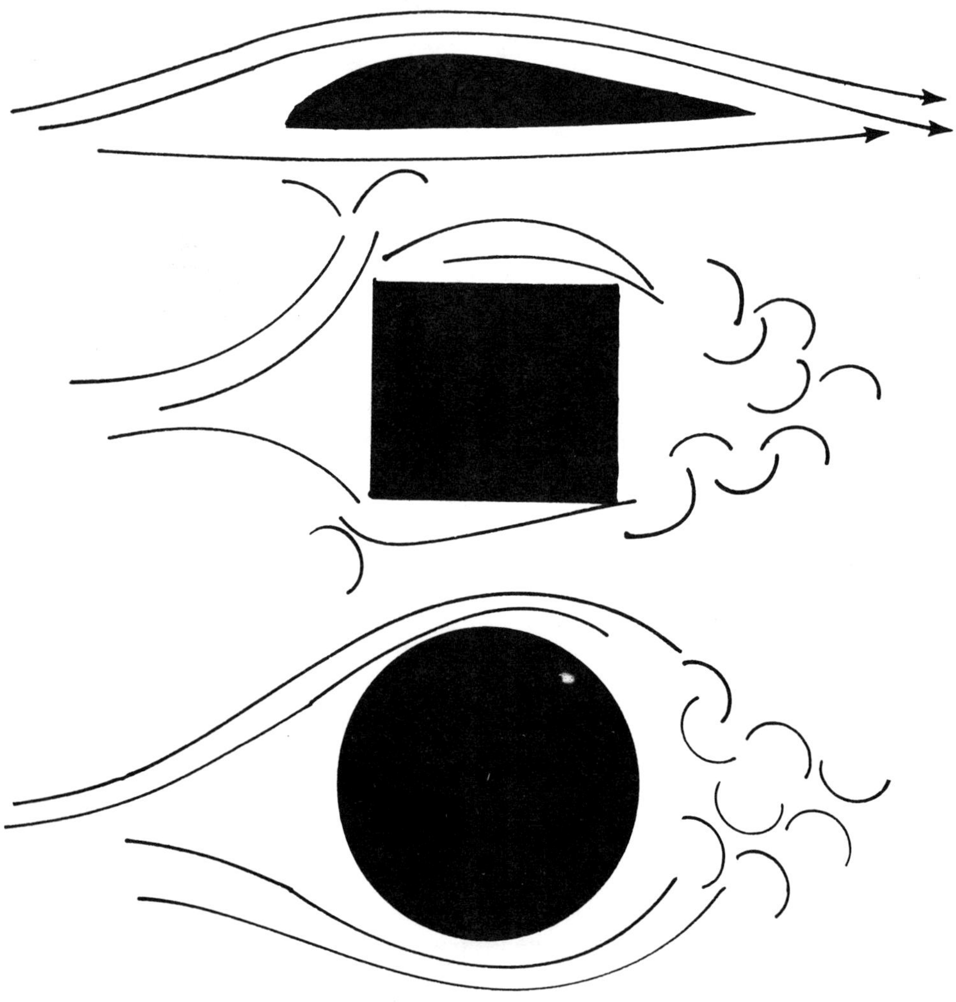

Drag will also be made when a model glider has not been sanded smoothly enough. The surface of the wood can be made very very smooth with fine sandpaper, the dope (paint for a model), and a thin coat of wax (paste floor wax is O.K.).

We used a simple homemade wind tunnel to try to see how the air flowed. We saw that most of the drag and turbulence was not on the Leading Edge, or the front of things, but on the Trailing Edge, or back of things. This was something we later read in books.

A vacuum sweeper was used to pull air through the tunnel. It seems easier to pull air than push it, because it doesn't create so much turbulence. We also found that it worked pretty well having one sweeper blowing air and one pulling air from the other end.

Dry ice in a cup of hot water was used to make the air easier to see.

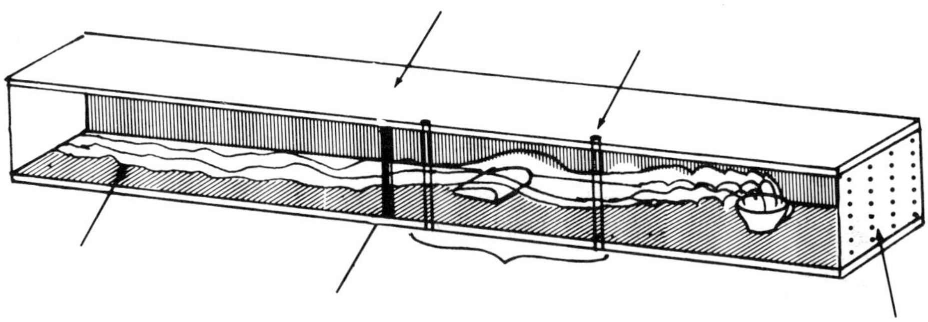

Materials:

½ pint of flat black paint (to paint the insides to see better)
A 1″ x 6″ x 8′ board (#3 pine O.K.)
3d nails
Two 1″ x 4″ x 8′ boards
3 panes of single strength glass: 5″ x 48″
5″ x 23″
5″ x 24″
Small scrap of pegboard 4″ x 5″ for the end to even out the flow of air.
Electricians tape to tape the glass on
Small scrap of wood about ¾″ x ¼″ x 5″ for center support
Rubber bands (to fasten shut the movable pane)
¾″ finish nails to nail on top and bottom to hook rubber bands on to
When glass gets cold it expands so leave some small spaces at the joints.

Wind tunnels are difficult to make because it is difficult to keep the turbulence down. This square cornered tunnel is not as effective as a tunnel with round sides, like if you used a series of Baskin-Robbins cartons it might work better. If you live near a University or College, you might check to see if there is a wind tunnel available for you to observe.

Although Drag is a force to be overcome, it is also necessary for controlling flight. This is how the control surfaces are used. A small amount of drag is put in one place to make it slow down there and the plane or glider can sort of pivot on that part.

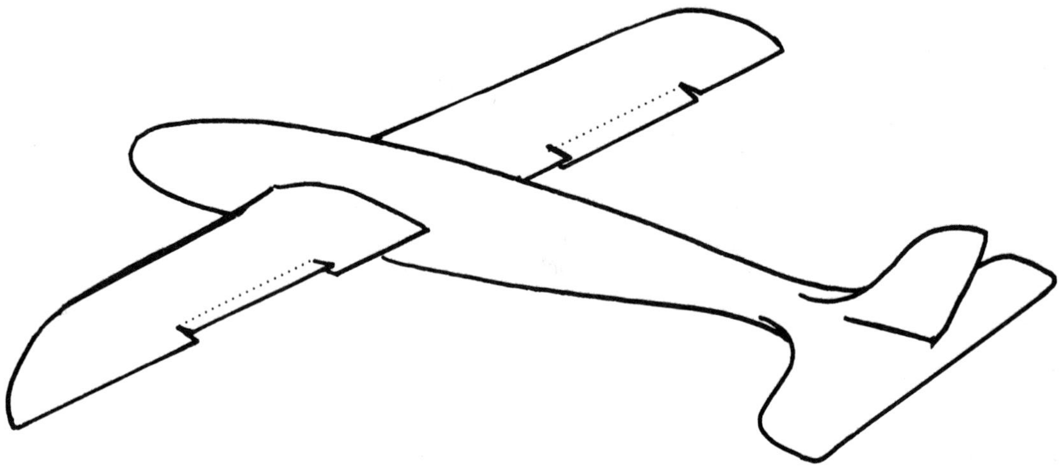

As long as there is more LIFT than DRAG the plane or glider will fly, but it will *STALL* if the *DRAG* is more than the *LIFT*. This will happen even if the airfoil and streamlining is done well, if the *ANGLE OF ATTACK* is too steep.

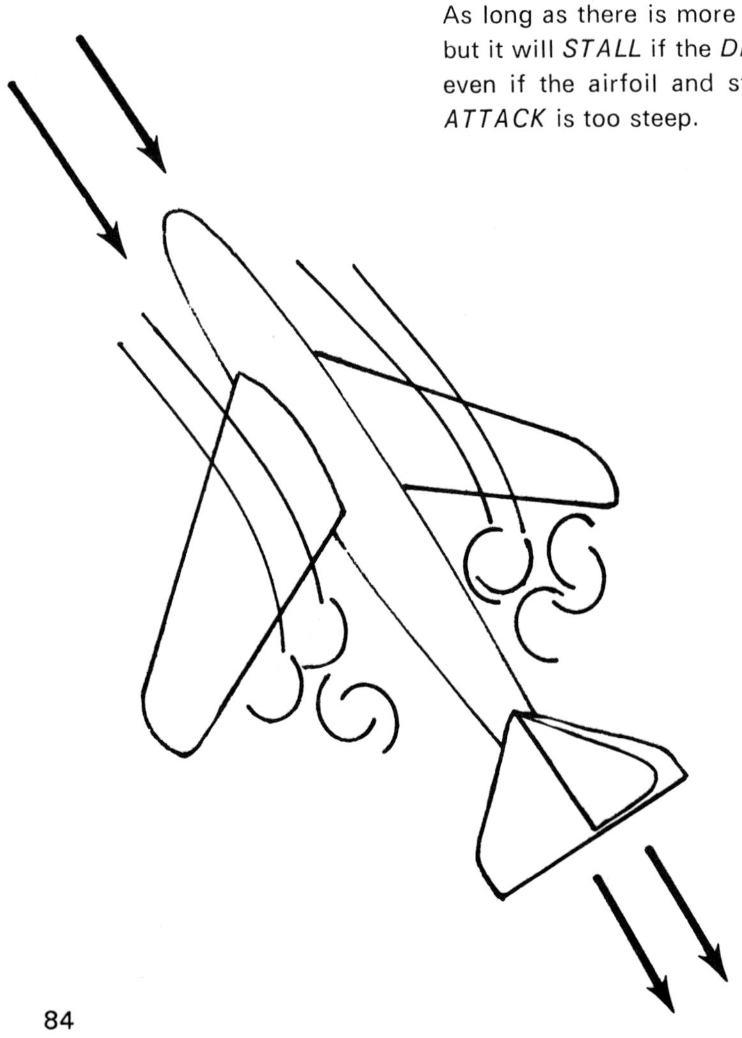

Some useful books for studying forces that affect planes and gliders are:

Barnaby, Captain Ralph, *How to Make and Fly Paper Airplanes,* New York, Four Winds Press, 1968

Corbett, Scott, *What Makes a Plane Fly,* Boston, Little Brown and Co., 1967

Gottlieb, W.P., *Aircraft and How they Work,* New York, Doubleday and Co., 1960

Herbert, Don and Ruchlis, Hy, *Beginning Science with Mr. Wizard,* New York, Doubleday and Co., 1960

Herbert, Don, *Mr. Wizard's Science Secrets,* New York, Hawthorn Books, Inc., 1952

Schneider and Ames, *Wings in Your Future,* New York, Harcourt Brace, 1955

Soaring Society of America, *Joy of Soaring,* 1969

Part Four

Development of the Glider Club

DEVELOPMENT OF THE GLIDER CLUB ....

After we built our first gliders we were excited and anxious to fly them. We took short trips to a field nearby where there were no trees or buildings that would make it difficult to fly our gliders.

We flew our gliders until they were all broken. Then we went back to school to fix them up.
We had so much fun, learning so many things, that we formed a Glider Club.

This was after we had been working on gliders for three or four weeks, on and off.

We made a bulletin board to show our gliders and tell things we had been learning. Also, we made notebooks. We did our lessons in them and answered questions, drew pictures, collected patterns and wrote about experiments we had done.

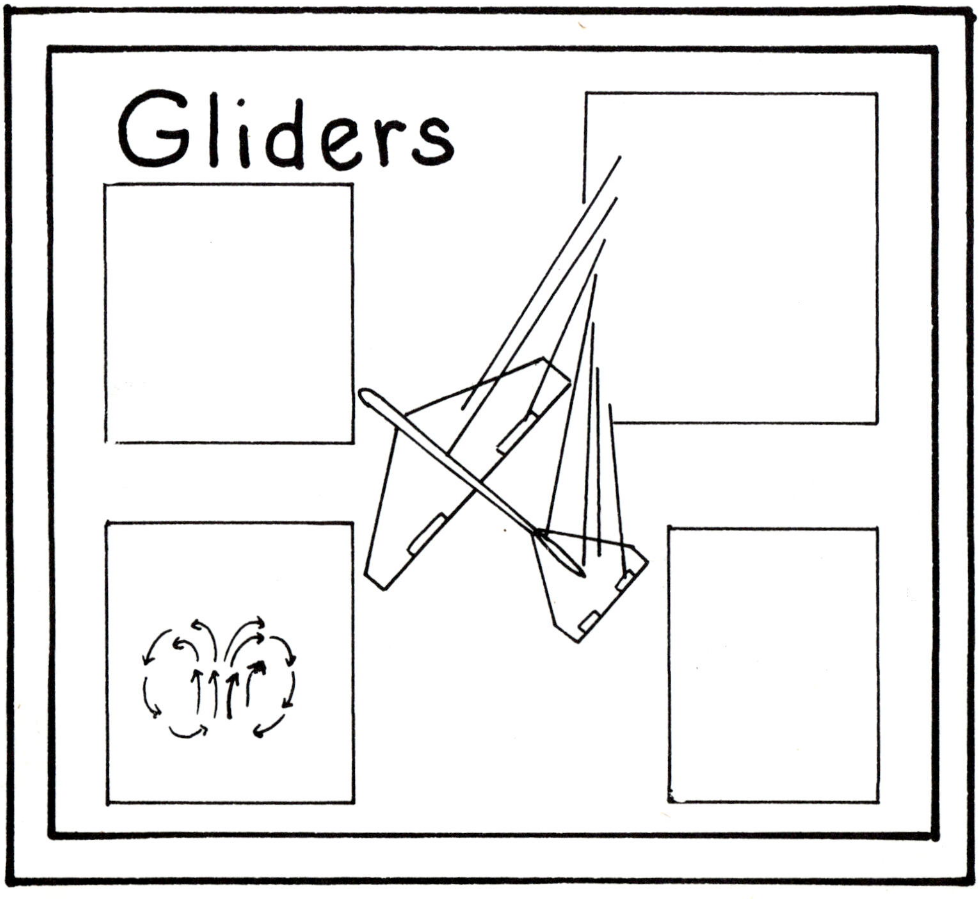

Dr. Bob Christopher from the University of Colorado came to help us. He showed us the best way to build a glider from a pattern and then he took us out to the field and showed us how to launch the gliders he brought with him.

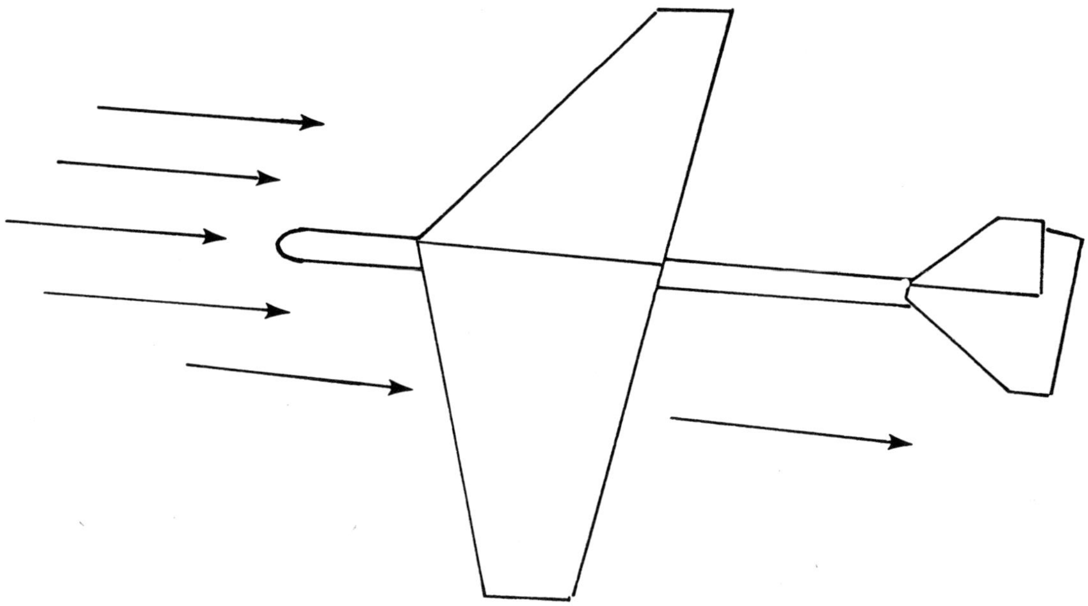

We wrote letters to Dr. Christopher thanking him for his help, for the patterns he gave us and for the *five* gliders he left for us! (What a surprise!)

We took field trips to the Boulder Glider Port, where we could see full size gliders take off and land.

Mr. Bruce Miller, the president of the Boulder Soaring Society met us at the Glider Port. He showed us how to work the control surfaces on two gliders. He showed us how the tow hitch worked. We saw a lot of people gliding. They would take off with a tow plane pulling them and when they were going fast enough and were high enough they would release the tow hitch. The tow line dropped and they would glide around very quietly and then come in for a landing on the landing strip. To go for a ride in a glider cost about ten dollars a person, so we didn't do it as a club activity, but a few of the kids in the club asked their parents for a ride and were given the chance to actually go up. Ted took some pictures when he was up and he recorded information about the trip and about the glider on his tape recorder when he got back!

After seeing full size gliders, we flew our own models in a field near the airport where it wouldn't interfere with air traffic.

One time when we went to the glider port, we saw a powered plane doing stunts! Sometimes our models did stunts and it was really fun to see a real stunt plane way up in the sky doing the same thing and other things even better.

We wrote letters to Mr. Miller telling him what we thought about the planes and thanking him for showing us things at the glider port. Bradley drew a picture and Danny wrote a poem.

On day for a lesson we answered some riddles about flying. Then we made up our own riddles to ask each other.

I go up he go's down The nose has agrate big frown. who am I.

The elevator

Danny

I don't like to fall... But it seems I always dive or I (stall.)

Gliders

Glidersare fast, Gliders are slow. Glider fly, Hi, Glider. Gliders fly low.

Danny

Two other very helpful people to go to for information about gliding and model building are Richard and Ed Whitten, authors of the *Star Skippers.* The *Star Skippers* was a national model plane newsletter for people interested in this same hobby: building and flying model gliders and planes.

The newsletter gave helpful information, like which models are good to build and will fly well: it also explained aerodynamic theory and tells the winners and all the entrants in the contests that they have 3 or 4 times a year. We entered a mini-contest (since our kids were pretty young.) After timing our flights with a stop-watch, we sent in the longest flight times for each glider. It was fun to see our names and times for the gliders printed in the next issue, even though our flight times were not very long.

The newsletter had an article about us becoming interested in hand-launched Gliders and they even printed up the cartoon that Chuck Hopper sent in!

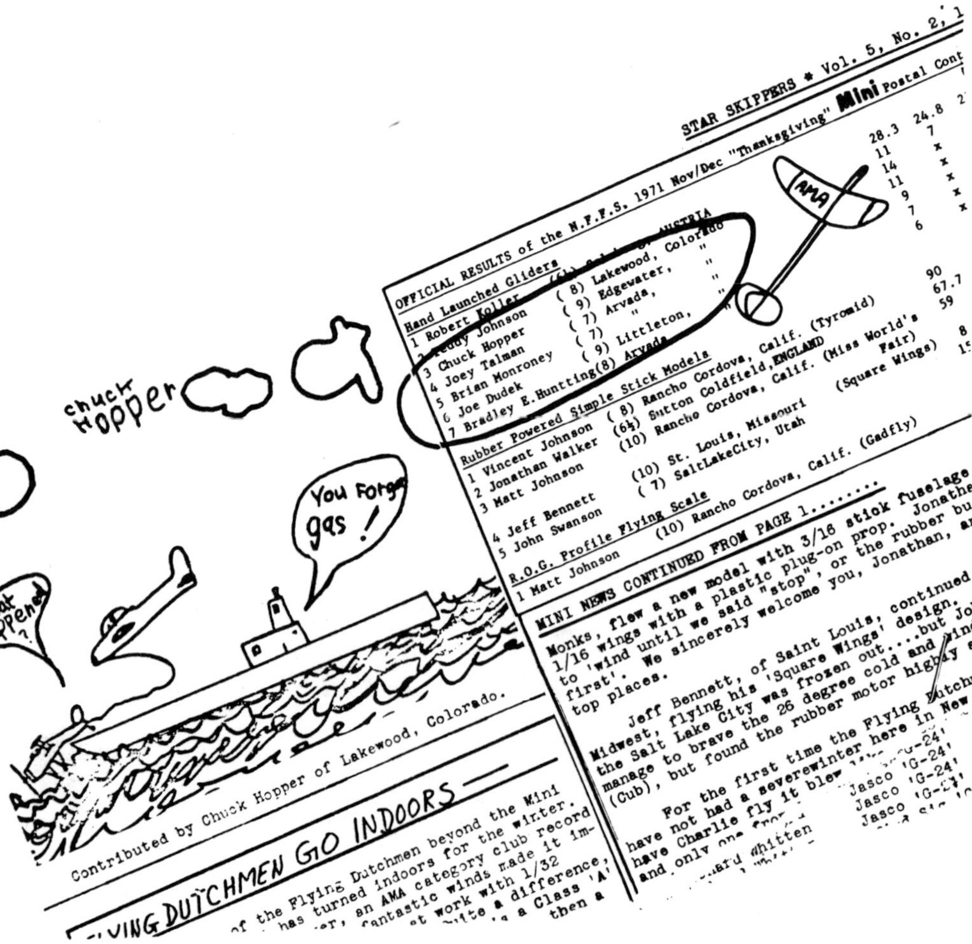

STAR SKIPPERS * Vol. 5, No. 2,

OFFICIAL RESULTS of the N.F.F.S. 1971 Nov/Dec "Thanksgiving" Mini Postal Cont

Hand Launched Gliders
1 Robert Koller (6½) AUSTRIA
2 Teddy Johnson (8) Lakewood, Colorado
3 Chuck Hopper (9) Edgewater, "
4 Joey Talman (7) Arvada, "
5 Brian Monroney (7) "
6 Joe Dudek (9) Littleton, "
7 Bradley E.Huntting(8) Arvada

Rubber Powered Simple Stick Models
1 Vincent Johnson (8) Rancho Cordova, Calif. (Tyromid)
2 Jonathan Walker (6½) Sutton Coldfield, ENGLAND
3 Matt Johnson (10) Rancho Cordova, Calif. (Miss World's Fair)
4 Jeff Bennett (10) St. Louis, Missouri (Square Wings)
5 John Swanson (7) SaltLakeCity, Utah

R.O.G. Profile Flying Scale
1 Matt Johnson (10) Rancho Cordova, Calif. (Gadfly)

MINI NEWS CONTINUED FROM PAGE 1.......

Monks, flew a new model with 3/16 stick fuselage 1/16 wings with a plastic plug-on prop. Jonathe to 'wind until we said "stop"', or the rubber bu first'. We sincerely welcome you, Jonathan, ar top places.

Jeff Bennett, of Saint Louis, continued Midwest, flying his 'Square Wings' design. the Salt Lake City was frozen out....but Jo manage to brave the 26 degree cold and wind (Cub), but found the rubber motor highly e

For the first time the Flying Dutchm have not had a severewinter here in New have Charlie fly it blew

Contributed by Chuck Hopper of Lakewood, Colorado.

FLYING DUTCHMEN GO INDOORS

of the Flying Dutchmen beyond the Mini has turned indoors for the winter. an AMA category club record fantastic winds made it im- at work with 1/32 quite a difference; a Class 'A' then a

We decided to have glider contest and we made posters to put up around the school each time we had a contest. Chuck and Brian took care of registration. The prize each time was an especially well-built glider . . . but if one person had won the prize glider once and then won again, he got a prize of balsa (to build his own) and the second place winner received the prize glider. That way no one got discouraged when one or two people got really expert at building.

Although we did a lot of building of models; flying them and adjusting them, we also wrote about them, and did a lot of measuring and figuring in our studies. But much of our time as a glider club was spent doing science experiments and talking about them.

Glider Contest					
1.					
2.					
3.					
4.					
5.					
6.					

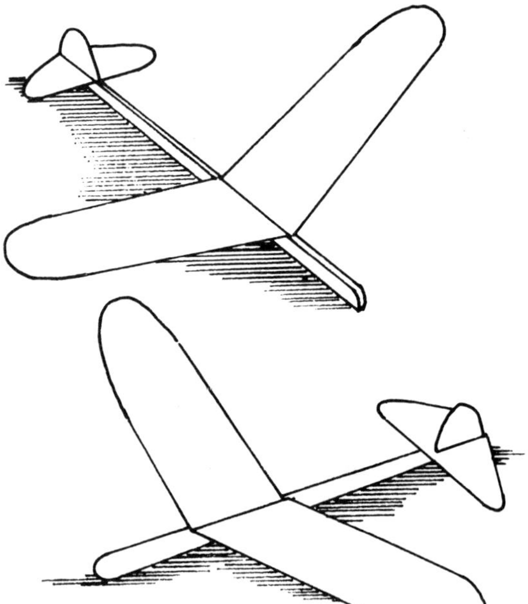

By the end of the year several children in the glider club had become experts. Some were experimenters, who tried out different ways to build planes. Some were exceptionally good at building planes from patterns. Most of them understood how to improve the flights of their model planes and how planes really flew. They had developed more of a feeling for air as a substance and understood more of how it worked and how to work with it.

These experts helped the newcomers to the club. They would teach the new members how to build planes that would fly better; how to balance the planes and how to correct the flights of the ones that didn't fly very well.

There was a particularly warm and comfortable spirit of camaraderie among the members of the glider club. A couple of very close friendships grew out of it and a general social growth took place, in that each member could empathize with others in times of discouragement. Many times one child or another would launch his carefully built glider for the first or second time, only to have it crash down and break. That feeling of discouragement was shared by others and there was almost always someone who would offer to temporarily replace the crashed plane with one of their own for a throw or two. And, too, the empathy with success and elation was similar. I think every member of the club was completely awe-struck as we stared wide-eyed at Danny's successful experimental plane on field day, as it went silently gliding, dipping and winging across the field. There were leaps and shrieks and gasps of excitement and joy. We all watched until it finally landed several minutes later, about a block and a half away! It was truly a beautiful flight which we all shared in, some of us having run halfway down the field in an effort to keep up with it!

The children changed; they grew much less competitive, although they did have contests and competitions. Somehow the moment of competition was reserved for the planes in the air, not generally for boys and girls to be against each other while working and building them. Many children became less isolated and more open to each other. Many of them pursuing their interest at home, brought balsa wood in to use and told of places they had found to buy balsa that was cheap but good. Their self esteem improved and their abilities and interests in academic subjects rose. The study of flight and aerodynamics, through the use of paper airplanes and balsa gliders was something which really touched a spark in them.

The learners became teachers and the teachers (myself for one!) became learners. The classroom responded positively to this interested, enthusiastic group of air students. And as a teacher, I found it the most rewarding study I've ever pursued with children.

by Glider Club **GLIDER News**

NEWTON'S LAW

SOMTHING GOING ONE WAY HAS AN EQUAL REACTION GOING THE OTHER WAY.

Who Was Berhoulli?

By Brenda

Air Going Fast can't Push hard.

air Pushes on all sides

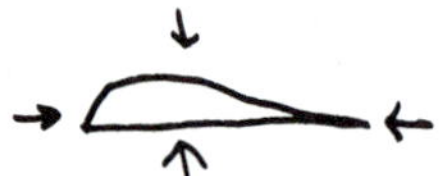

fast air can't push down so wing goes up.

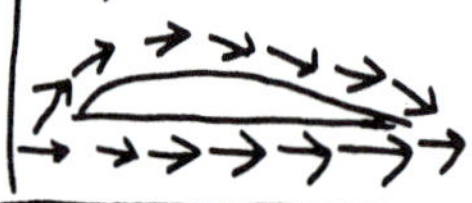

A GLIDER RUSHING ONE WAY CREATES WIND THE OTHER WAY

BY BRIAN

glider Contest!

Build your own glider
No Precut kits
only Hand Launched gliders

by Chad

April 12 1:00 meet in Judys room

lift

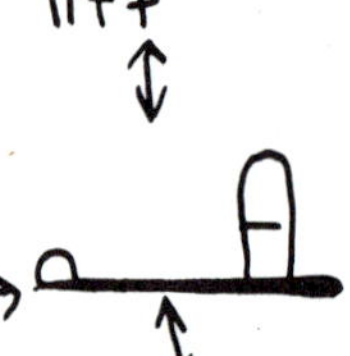

There are four forses on a glider: lift Drag, gravity, and thrust. lift pulls up, Drag pulls Back, gravity pulls down, thrust pulls forword

by Danny

by Stanford

If The ailerons are Like this the glider will turn rigt. if the rudder is turned left the Plane will Turn Left

If The elevators are up on the left and down on the rigt IT goes left.

OTHER SUBJECTS OF INTEREST PRODUCED BY A STUDY OF GLIDERS

Powered Aircraft including Jets and Rockets
This can be a natural subject to follow through on when studying gliders. Many children of ages 8 to 11 are particularly interested in it.

Balloons and Lighter-Than-Air Craft
This is also a natural topic that comes up in the subject. It is easy to follow through on as a topic for younger children, as well as older ones.

Birds
The original gliders are of course fascinating and beautiful and many children are interested in learning to identify them, make bird houses and learn some of the songs.

History and Museums
Learning the history of flight can spark an interest in making museum displays at school, which man spark interest in visiting museums and in exploring history in other subjects that happened at the same time.

Balsa Modeling and Sculpturing
This art project is rewarding and involving and can be done with the scraps left over from building. Boats, autos and sculptures were the most popular things made in our group.

Sink and Float
A science study about weight, sinking and floating in water, or oil, or salt water, etc. can be appropriately introduced. It was suggested by a couple of children, during our glider study.

CO_2 Experiments
Having the dry ice around for the wind tunnel and other experiments can lead to questions and interest in the ice itself. Putting out a flame; testing the gas, plants give off, to see if it's the same and other experiments, are exciting to follow up on.

Newspaper Writing
Class newspapers are nothing new, and are still fun for a lot of kids. The idea of fliers, and news bulletins or single sheet broadsides came to us when we did our own bulletin on gliders. A one leaf flier on a single topic can sometimes be more appropriately included in a study of something.

GLOSSARY

AILERON . . . . the hinged part of the Trailing Edge of the wing near the wing-tip. The ailerons are used to make the plane roll to the right or left. This helps it turn.

AIRFOIL . . . . any part of a plane (mainly the wings) which deflects air in such a way that the air will LIFT the plane or DRAG it, depending on what is intended.

ANGLE OF ATTACK . . . . . the angle at which the wing hits the wind. This has nothing to do with the horizon.

BANK . . . . the plane tips or rolls to the side. This helps it turn.

CAMBER . . . . the cross-sectional curve of the wing.

CENTER OF GRAVITY . . . . the balancing point of a glider or plane. (abbr. C.G.)

CENTER OF LIFT . . . . the point where the forces lifting a plane are balanced. (abbr. C.L.)

DIVE . . . . the plane heads down nose first.

DRAG . . . . the force opposing thrust. Drag pulls a plane back.

ELEVATOR . . . . the horizontal movable surface on the tail of a glider or plane.

EMPENNAGE . . . . the entire group of tail attachments on a plane. It includes moveable and unmovable parts.

FIN . . . . the fixed vertical tail piece.

FLAPS. . . . . hinged parts of the Trailing Edges of both wings, between the ailerons and the fuselange. They are used to change the LIFT and DRAG of the wing.

GLIDE. . . . . forward flying, speed kept up by going down. (A glider gains altitude by going *down* in upward moving air which is going up faster than the glider is going down.)

GLIDER. . . . . any wing supported craft with no power of its own. It is the same as a sailplane but a sailplane is used most often to mean a plane that soars cross-country instead of just taking brief downward glides.

GLIDE RATIO. . . . . a comparison of how far a glider travels forward to how far it travels down.

LEADING EDGE. . . . . the front edge of an airfoil (wing). (abbr. L.E.)

LIFT. . . . . the force which makes a plane go up. Also, upward flowing air (like a thermal) can be considered lift.

PITCH. . . . . plane nose up or down: movement of the plane on a horizontal axis going across the fuselage.

RUDDER. . . . . the hinged part of the vertical tail. The rudder helps make the plane turn left or right.

ROLL. . . . . the wing tips move up or down: movement of the plane around a horizontal axis going parallel to the fuselage.

SINK. . . . . down-drafts: or downward air currents which make a plane go down faster than if it were in still air.

SOAR . . . . a plane that is flying without engine power and is not losing altitude, and is usually going on longer distance flights.

SPOILERS . . . . things like flaps above and below the wings to spoil the LIFT and increase the DRAG.

STABILITY . . . . a glider has stability when it flies straight, in a steady glide, without the pilot doing anything. It also has stability if it returns to the glide after being disturbed!

STABILIZER . . . . the fixed horizontal tail surface, or vertical tail surface, either one.

STALL . . . . a sudden total loss of LIFT.

THERMAL . . . . rising warm air that has been heated by an underlying surface. It is surrounded by cooler air.

TRAILING EDGE . . . . the back edge of an airfoil. (abbr. T.E.)

YAW . . . . the nose of the plane moves to either side: movement of the plane around a vertical axis perpendicular to the fuselage.

Em
255003

508 9